VIRAL FITNESS

Also by Jaap Goudsmit: *Viral Sex: the Nature of AIDS* (1997)

The publication of this book has been made possible with financial support from the Foundation for the Production and Translation of Dutch Literature.

VIRAL FITNESS

THE NEXT SARS AND WEST NILE IN THE MAKING

Jaap Goudsmit

OXFORD
UNIVERSITY PRESS

2004

OXFORD
UNIVERSITY PRESS

Oxford   New York

Auckland   Bangkok   Buenos Aires   Cape Town   Chennai
Dar es Salaam   Delhi   Hong Kong   Istanbul   Karachi   Kolkata
Kuala Lumpur   Madrid   Melbourne   Mexico City   Mumbai   Nairobi
São Paulo   Shanghai   Taipei   Tokyo   Toronto

Copyright © 2004 by Oxford University Press, Inc.

Published by Oxford University Press, Inc.
198 Madison Avenue, New York, New York 10016

www.oup.com

Oxford is a registered trademark of Oxford University Press

Library of Congress Cataloging-in-Publication Data
Goudsmit, Jaap, 1951–
Viral fitness : the next SARS and West Nile in the making /
Jaap Goudsmit.
p. cm.
Includes bibliographical references and index.
ISBN 0-19-513034-0
1. Viruses.   2. Virus diseases.   I. Title.
QR201.V55 G68 2004
614.5'8 — dc22   2003021606

1   3   5   7   9   8   6   4   2

Printed in the United States of America
on acid-free paper

To Carleton D. Gajdusek

PREFACE

The September 11, 2001, attack on New York and Washington, D.C. changed not only the world of humans but also the world of viruses. No one would have thought it possible that a group of terrorists could strike the United States at its very heart, yet that is what happened when the twin towers were destroyed. Many people now urgently believe that we must arm ourselves against a repeat of that attack, especially if it should come in the form of bioterrorism.

The smallpox virus has been eradicated, Ebola virus infection is very rare, and America and Europe seem to be free of foot-and-mouth disease. But we must now live with the fear that, one day, someone will breathe new life into these or other viruses. Since September 11, we must consider that a terrorist would intentionally spread a virus such as smallpox or Ebola among people or introduce the agents of food-and-mouth disease, hog cholera, or mad cow disease into the food chain. For more than twenty years, we have lived with the belief that smallpox would never return; nobody would ever again have to be vaccinated against that terrible disease. But since September 11, smallpox vaccine is again in production. Initially it is for the entire population of the western world, the focus of terrorists, but worldwide coverage could some day be necessary.

There used to be two different worlds of viruses. There were the viruses in the wild, which were making their way through nature in accordance with evolutionary laws. And there were the vaccine viruses:

strains manipulated by humans to provide defense against their more aggressive relatives. Vaccine viruses protect us not only from smallpox but from polio and measles, for example. Now a third world of viruses has been added: strains manipulated by humans to attack other humans. Thus a new biological and evolutionary reality has been created, the reality of viruses as poison.

In fact, the word *virus* means poison in Latin. Viruses in nature sometimes live up to their lethal reputation, but often they do not. When they were discovered and named early in the twentieth century, they were mainly plant viruses. These first viruses were considered pests, causing plant and animal diseases that sentenced farmers to poverty and could cause thousands of people to die of starvation.

Nowadays, especially in affluent countries, we pay far less attention to plant viruses than to human viruses. And although HIV can ruin a life, we think mostly of viruses that spoil only a day or a week. For most of us, they are a short-term problem, but at the same time, we all know consciously or unconsciously that there is no life without viruses. As children, we almost inevitably have our bout with chicken pox. In our teens, we may get mononucleosis and, during adulthood, there are cold sores and shingles. In addition to all this, we have colds and perhaps the annual flu. We have antibacterial drugs but very few antiviral drugs, so a viral illness usually runs its course and we must simply wait to feel better.

As this book will show, some viruses are more avoidable than others. But every day, millions of children get diarrhea or red spots due to a virus. Every day, millions of people turn yellow or green due to a virus. At any given time, there are more people walking the earth with a virus infection than without one. Indeed, we all carry some virus or other throughout our life. Fortunately, most of these harm us infrequently, if at all. They multiply in silence, without our being in any way aware of those millions of virus particles which we breathe out, spread by sneezing, or transmit to our fellow humans through sexual intercourse.

Unfortunately, there are a few nasty viruses that seem to be unable to propagate without causing disease, and we can be sure that many of us will die from such a viral disease, whether it is AIDS, liver cancer, or leukemia. People in Asia and Africa, where life is already hard enough, are even more likely to die of these viral diseases and many others.

There is no such thing as life without viruses, neither among people nor among animals, plants, or bacteria. Being a nuisance to us: is that really the only thing viruses do? If so, we should get rid of them as quickly as possible! But the very idea of getting rid of all viruses should be abandoned. There is no chance whatsoever that we will succeed in eradicating all viruses. There are simply too many different kinds, and they all multiply much faster than any animal or other host. Within a few hours of infecting their favorite host cell, viruses use cellular machinery to produce millions of offspring, which in turn penetrate new cells. Before we know it, our blood and other bodily fluids are teeming with virus particles.

Despite vaccines and medicines, we have so far failed to banish many threatening viruses, including those that cause jaundice, measles, polio, many kinds of diarrhea, and of course AIDS. We have succeeded only in eradicating one single virus, smallpox. We could do this because it is a virus that occurs only among people; it could not jump to another animal for survival. Moreover, it causes an infection that can be recognized from the telltale facial lesions every time it strikes. Most viruses are harder to detect and fight, because they spread without any visible signs. Also, they can hide most of the time in chickens, ducks, pigs, horses, and rats, and then suddenly show up among people.

Viruses appear to be almost immortal, while species of animals and plants are becoming extinct almost daily. If survival is for the fittest, viruses seem the fittest of all creatures in their amazing adaptability. What is it that makes them so incredibly strong—and can we learn something from it? What viruses can we control or avoid? Will viruses spell the end of the human race, or will we always be able to offer resistance?

That is what this book is all about.

ACKNOWLEDGMENTS

In addition to many invaluable people, specific moments and circumstances have contributed to the creation of this book. Without those people and events, I would not have undertaken or completed the book. Nor would it have become the book that now lies before you.

The idea began to germinate in 1994, ten years after I had started my research on AIDS and shortly after I had begun to write *Viral Sex: The Nature of AIDS*. In spring of that year, I attended a meeting organized by the Rockefeller Foundation in Bellagio, Italy, where participants came to the conclusion that the only "real" solution to the AIDS problem was a vaccine. What this vaccine would be like and how it should be made was discussed in fall 1994 at a gathering in Paris. The people who sat around the table were scientists who have all shared thoughts that have helped to shape this book. Marc Girard guided me to the conviction that vaccines are not only the best weapon against HIV but against all viruses. Andrew McMichael and Rolf Zinkernagel made me aware of the complexity of our battle against viruses. Esteban Domingo kept me on the right evolutionary track.

The International Aids Vaccine Initiative (IAVI) was founded in 1996 to get an AIDS vaccine to the people in need. Contacts with the IAVI crowd and in particular with its CEO Seth Berkley convinced me that virology is not only a science, but also the basis for saving people from viral assaults. Over meals in great cities around the world, I also talked many times with Neal Nathanson about the newest developments in virology. Neal is a great virologist and was director of the National Institute of Health's Office of AIDS Research and later a member of IAVI's scientific advisory committee, that I chaired from 1996 till 2002.

Once I had joined the IAVI board in 2002, I met Richard Feachem—the person who, more than anyone else, made me realize the importance of public health. Another influence was Barry Bloom, who convinced me that fundamental research, epidemiology, clinical research and public health must be inextricably linked if we are to conquer the ever-increasing number of infectious diseases. He was also the first person to make me aware of the relationship between poverty and infectious diseases. This awareness led me to start the Center for Poverty Related Communicable Diseases, a think tank at the Amsterdam Medical Center.

In September of 2001, this book gained an additional dimension. On Saturday, September 8, I attended a birthday party for Laurie Garrett, author of *The Coming Plague*, at the Brooklyn Brewery. Wayne Koff, the science director of IAVI, his wife, and I shared a cab to Brooklyn from IAVI's office in the Wall Street area. Totally unaware of the disaster that would soon occur, we celebrated a person whose work has inspired me time and again. The following Tuesday, September 11, at half past eight in the morning, I arrived at an IAVI meeting to evaluate vaccine programs. Located on the twenty-seventh floor of 110 William Street, the conference room offered a clear view of the World Trade Center three blocks away. At a quarter to nine, one of the IAVI employees called us to the window, saying that an airplane had hit the north tower. We looked at the fire in amazement. It hardly seemed real, but as we stood there twenty minutes later, we saw with our own eyes a second airplane flying into the south tower. Everyone knew at that moment that this tragedy would change the world.

A few minutes later, we were standing with all the IAVI employees and thousands of other people on the William Street sidewalk. Finally, we could not think of anything better than to go to our hotel on nearby Gold Street and continue our meeting. That idea turned out to be ridiculous when, at ten o'clock, the first tower collapsed and, half an hour later, the second. Many people who had escaped from the towers came to that hotel to recover and tell their stories.

With towels held to our faces against the thick white smoke, Ian Gust and I went to the South Street Seaport, where we were able to breathe better due to proximity of the water. There we talked while watching a crew of tugboats that had organized to offer free transportation to the many people in need. Of all my inspiring conversations with Ian, I am perhaps most thankful for that one by the water in Manhattan. By evening, all the participants in the IAVI meeting were walking in silence among thousands of New Yorkers, all making their way uptown along South Street to safer regions.

From that day on, nothing has been unthinkable anymore, not even an attack with smallpox or Ebola. Writing this book became even more important to me, and many more people gave their time, help, and good will to the project. I want to thank Erna Albers and Carina Hastrich for the countless hours that they dedicated to preparing the finally completed manuscript. I greatly appreciate the work of Karin Ford-Treep, who translated the early Dutch draft of this book.

This book would never have happened without Lucy Phillips, my executive editor. Lucy kept me on the critical path and helped to shape the book. Absolutely nobody deserves more credit for *Viral Fitness* than she. I thank my editor at Oxford University Press, Kirk Jensen, for his continuous support to the project.

My wife Fransje has helped me to bring this book to a happy conclusion, "because I wanted it so much." I also thank my three splendid daughters, Judith, Leah, and Keziah, for the time which was really theirs and which they had to give up for the writing of this book.

CONTENTS

Introduction: Virus Basics and the Concept of Viral Fitness   3

1   As Natural As Breathing: The Flu Virus   13

2   Farming and Feeding the Hungry: Plant Viruses
    and Human Enteroviruses   26

3   Raising Cattle and Eating Meat: Rinderpest, Measles,
    and Mad Cow Disease   41

4   Slaking Our Thirst: The Cholera Bacteria and Its Toxic Viruses   56

5   Weathering Storms and Droughts: West Nile Virus and Others   67

6   Getting Lucky With a Faulty Gene: Escape from
    Simple Retroviruses   79

7   Taking Chances With Sex: The Herpes and Papova Viruses   93

8   Risking Death with Sex: The AIDS Virus   108

9   Warring Against Humans and Other Animals:
    Smallpox, Monkeypox, and Others   125

10 Raiding the Wild For Delicacies: The SARS Virus   140

Epilogue   153

Glossary   163

Bibliography   171

Index   185

VIRAL FITNESS

INTRODUCTION

Virus basics and the concept of Viral Fitness

*R*arely have humans been as threatened by viruses as they are today. It almost seems as if a virus invasion is taking place. Viruses have lately been coming out of nowhere and appearing in the strangest places— exotic viruses about which no one had ever heard before. The West Nile virus shows up in the New York borough of the Bronx, the Hendra virus kills a horse trainer in Australia, the Nipah virus is the cause of encepha- litis in Malaysia. A mysterious influenza virus infects children in Hong Kong, the Hantaan virus attacks Navajos in New Mexico. And finally a new coronavirus kills people in China, and a monkeypox virus, never seen before outside of Africa, shows up in several states of the United States. All these outbreaks have occurred since HIV launched the AIDS epidemic in the early 1980s.

Many human viruses have started out in the animal world. House sparrows, starlings and crows in the Bronx are carriers of the West Nile virus, the Hendra virus is a horse virus, the Nipah virus is a hog virus, the Hong Kong flu virus is a chicken virus, and the Hantaan virus is a virus of rodents. Finally, the AIDS virus is a simian virus and the SARS virus is a virus of the civet cat. Are these viruses maybe escaping from their nat- ural hosts? Are human beings simply available as easy prey? Why is all this happening now—and what does it mean for our future? What can we do to defend ourselves?

In this book, we will try to find answers to these questions. But first, what exactly are viruses and how do they fit into the living world?

The living world includes three enormous families: bacteria, archaea, and eukarya. Everybody has heard of bacteria, but what about the other two? The archaea are very primitive one-celled organisms, like bacteria, which thrive under extreme conditions that recall the earliest times on this planet. The eukarya are cells with a nucleus, including single-cell organisms like amoeba and paramecia and all the cells of all the multi-celled organisms on earth: all plants and all animals, from insects to humans. Viruses are parasites in this living world. Since their discovery in the early twentieth century, there has been ongoing debate as to whether they are actually alive.

A living being is an independent entity that is clearly separated from its surroundings. Since the beginning of life on earth, its basic unit has been the cell. It can be the one and only cell of a one-celled organism, or it can be just one of the millions that make up a plant or animal. Either way, each cell is an independent and separate life form. Its mitochondria and other organelles are as busy as our heart or kidneys. Along with the cell nucleus, these bodies are suspended in the nourishing cytoplasm, which is enclosed by a wall or membrane. The wall is itself an active organ, letting things in through receptors, sending signals, and performing other duties much like our skin. In the nucleus is stored a kind of computer program that determines all the cell's characteristics and behaviors, such as being able to eat, drink, breathe and produce offspring. This software is structured to copy itself and consists of DNA or RNA, in the living world and in viruses.

Viruses meet every requirement of "living being," up to a point. Being smaller than bacteria, they cannot be seen under a light microscope, but under an electron microscope a virus seems like a separate particle with an obvious shape. The shape differs per virus family but is the same for all of that family's members. Similarly, all humans look like humans, as opposed to other animals—and humans of certain groups are recognizably members of those groups while still being part of the overall human race.

The virus particle is enclosed in a protein coat, or shell. This is less of an active organ like the cell membrane, although it has an inner wall and an envelope, or outer wall (see chapter 6). The virus interior, too, is simpler than what we find in a cell, although it has something like a nucleus that contains RNA or DNA code. As in living things, this code determines all virus characteristics and behaviors—with one crucial difference. The code is not quite complete. It contains all the information needed to assemble a new virus particle but cannot handle basic reproduction.

Every virus, in order to multiply and spread, must enter a living cell —in a host ranging from one-celled bacteria to multicelled plants or animals. It uses some of the cell machinery to produce its offspring. The act of cell entry or penetration is called infection, but it may or may not cause disease. Many infections, whether viral or bacterial, go unnoticed.

So a virus is not completely independent and therefore is not completely alive. Viruses are a kind of transitional form between the biological world and the physical or chemical world. They are just short of being alive. It is clear, however, that a virus particle only needs a small push to be brought to life. And if a virus exists, even in a dormant state, it must once have reproduced—in a cell—and if it gets the opportunity, it can reproduce again.

Viruses never really die. If viruses are not actively reproducing, they can indefinitely maintain an inert state. Every decade, every year, every month some creatures of the living world become extinct, but not so with viruses. Paradoxically viruses do not die because they are not alive. Viruses can be completely built or re-built using their genetic code. If we were to keep freeze-dried virus particles in an ampule for hundreds of years, we could still bring them to life. Under an electron microscope, they would still look like viruses and would immediately wake up if brought in contact with a living cell. For a virus, just its RNA or DNA code is enough for survival. All we have to do is place this genetic material inside a living cell, and the virus will spread again.

Individual viruses can be killed, as when radiation destroys the RNA or DNA code or extreme heat makes the virus particle disintegrate. But viruses are hard to kill because of their protein coat and their never-quite-alive quality. In addition, a virus population is protected by being able to evolve very quickly. Like bacteria, viruses multiply so fast, so often, and so much that, at any point in time, a virus population is absolutely enormous. It is also highly varied due to the mutations that naturally occur. The population still represents one virus family, but its members are so many and so varied that nothing we do to kill them will entirely succeed.

If you have a virus infection and take a drug to kill the viruses, a few will always survive no matter how many millions are killed. They will survive because they happen to be family freaks with a chance mutation that makes them immune to your drug. Suddenly these freaks have no competition from brothers and sisters. The majority has vanished, so in a few reproduction cycles over maybe 24 hours, the freaks become the millions of a new majority. And all have inherited immunity to your drug—on top of some with the inevitable mutations that prepare them for the next attack from the environment.

This virus knack for survival—the variability that makes them so fit—is a big reason why pharmaceutical companies have developed so few antiviral drugs. They simply do not work nearly as well or as long as antibacterial drugs work against bacteria. Even the antibacterials rarely work forever, since bacteria, like viruses, can evolve into resistant populations. Viruses just do it much, much faster.

All life forms can be divided into species, but viruses cannot. The variability and lightning-quick evolution of viruses brings us to another reason why they are not quite like living things. They cannot be separated into species because they change so readily. What is a species? As an everyday word it simply means type or kind, but in biology it is an official identification category. Every living thing is classified from its most general form to its most specific form: superkingdom (bacteria, archaea or eukaria) to genus and species. For example, humans are Superkingdom— Eukaria; Kingdom—Animalia; Phylum—Chordata; Class—Mammalia; Order—Primates; Family—Hominidae; Genus—Homo; Species Sapiens. The language of classification is Latin, but with many Greek roots.

Usually scientists refer to a living creature by its genus and species names together, in italics, so we are *Homo sapiens*. Bacteria and archaea are similarly classified, so we have *Escherichia coli*, a generally harmless bacterium in the human gut, and we have *Helicobacter pylori*, a bacterium that can give us stomach ulcers. After such genus and species names are first mentioned in a text, they can be curtailed to *E. coli* or *H. pylori*. The names may honor a person, as in "Escherich's bacterium of the colon" or describe the creature in some way, as in "helical bacterium of the pyloric valve."

Living things can be classified and specified because they are relatively stable in form for millions of years, evolving very slowly and mating only with their own kind. Creatures belong to the same species if they can maintain the species by bringing new members into the world by reproduction. We all know that humans can only beget human offspring. Humans cannot mate with non-human primates, for example. Most plants and animals reproduce sexually, by joining male with female; bacteria reproduce asexually, by cell division. But even bacteria do not mix species with species.

Such stability has advantages for living things, but since their evolution is slow and interspecies mixing is impossible, stability can lead to extinction. This happens if the environment changes too much or too fast for a species to evolve adaptations. For example, gorillas are now highly threatened and cannot enlarge their numbers by mating with apes of another species.

Viruses can be distinguished from each other genetically. Viruses do not lend themselves to speciation but can be distinguished in various ways, at least in the short term. Some have a single-string RNA or DNA code; others have a double-string RNA or DNA code. The sequence of nucleic acids in the DNA or RNA of a virus determines the place of a virus in the genealogy of all viruses. In addition, they can be distinguished on the basis of whatever living creature they infect. They can sometimes be distinguished by the type of cell they infect within their chosen host or hosts. But many viruses can infect both animals and humans. And there even are viruses that can infect plants and animals. This versatility comes not only from their quick evolution but from recombination and reassortment — a process of "interspecies mixing" that is impossible in the living world.

Even very different viruses can easily exchange genetic information if two conditions are met. First, the potential mates must simultaneously infect the same cell, which is not so improbable, given the vast numbers involved. Second, they must have a minimal piece of RNA or DNA in common. Recombination is especially easy for viruses that use RNA as their code. In fact, viruses with code in double-string DNA recombine less and evolve much more slowly than RNA viruses. They behave — and can be classified — much like creatures in the living world, who also have double-string DNA.

When virologists talk about divergent, rapidly changing RNA virus families, they sometimes use the word *quasi-species*, a term that was created by Manfred Eigen. Since viruses are almost-alive, it seems right that they should form almost-species. They are not a real species because their constant mutation and recombination or reassortment makes them adapt to changing circumstances with lightning speed. There are instances of viruses converting themselves in no time from a true bird virus to a human virus by way of a hog virus. There even are viruses that exchange genetic material with the host cell in which they multiply. And of course there is HIV, a special type of RNA virus, which keeps evolving resistance to drug therapy.

Maybe viruses even have eternal life. Gorillas are currently dying out, as are many other simian species. Maybe all life forms will become extinct in the long run—except for bacteria, which are protected by their enormous diversity. Maybe new life forms will again arise on the earth. In such a scenario, what would happen to the viruses? When animals or plants become extinct, are their viruses lost with them?

It is often assumed that each life form has its own viruses. If this were strictly the case, dinosaur viruses would have died out with the dinosaurs. But there is nothing to indicate that this is what happened. Probably dinosaur viruses jumped to unrelated animals or are now found in dinosaur descendants among birds and reptiles. The fact is, viruses will always find a way to survive. They mix with other types of viruses, they jump over to a different kind of host, or they go into hiding in their host's DNA.

Some viruses are fitter than others. What Charles Darwin called "natural selection, or the preservation of favoured races in the struggle for life," was called "survival of the fittest" by a 19th century contemporary, Herbert Spencer. This is the kind of fitness meant by the title of this book. The fittest viruses are those that most readily adapt to the environment. Virologists consider a specific virus population (or individual) fitter than another when it is better able than another to multiply and spread, given certain conditions.

Which virus is fitter changes as fast as conditions change. And while virus populations compete with one another, there is also individual competition within those populations. In fact, with viruses, group fitness and survival can hardly be separated from individual fitness and survival. Among living species, a population can include individuals that are more and less successful. Those that succeed do not blot out those less successful, in most cases. And if a few individuals die, the population still thrives.

But with viruses, the fittest individuals not only thrive more than the others: they soon overgrow the others. The less fit individuals virtually die out. However, those fittest under one set of conditions can quickly be supplanted if a new set of conditions suddenly prevails—as when an HIV-infected person changes his inner environment by taking antiretroviral drugs. The virus that comes out on top might have been barely surviving under the old circumstances, but now it is in the right place at the right time.

Viruses that are not fit do not survive. Clearly, this is not the same kind of "fitness" for which humans nowadays go to their local gym. For them, "fitness" involves being physically healthy and attractive, and, although such fitness can help with survival, it is not really about adaptation and selection. In affluent parts of the world, people have too many options to need that kind of rock-bottom fitness; they are too insulated. But fitness for survival was the basic rule for our cave-dwelling ancestors. It is still the rule for humans who live in such poverty or turmoil that survival is a daily issue. It is the rule for animals in the food chain. Predators and the animals on which they prey are always fit—or they are dead. When predators are no longer fit enough to catch prey, they starve. When prey are no longer fit, they are caught and eaten.

The virus fitness champions: bacteriophages and retroviruses. These two groups of viruses are the most adaptable to changing conditions because, due to options or insulation, they are the most protected from such conditions.

Bacteriophages are the viruses that infect bacteria. Nicknamed phages, they have the most options because bacteria are the most varied life form on earth. If phages ever need a new host, there are many to jump to. Phages rarely have to do much to survive and, in fact, often what they do is helpful to their host. They can even be helpful to humans (see chapter 4).

The other fitness champion is the retrovirus. It enters its chosen host cell like any other virus but can subsequently splice itself into the genetic material in the cell nucleus. It is then insulated and gets a free ride, surviving as long as the host survives. If a retrovirus just happens to infect a very early embryo (before the embryonic cells begin to differentiate into many types), it will wind up spliced into every single cell of the body, including the eggs or sperm. Thus it will be carried forth to all future generations of the host, surviving as long as the host species survives.

Retroviruses infect a more diverse collection of life forms than any other virus, except for phages. They are mostly harmless and were almost completely unknown until the emergence of HIV, the black sheep of the family. The retrovirus can transcribe RNA code into DNA, once thought impossible. This trick lets it slip into the host genetic material, which may or may not harm the host. The trick also may lie at the very basis of life on earth and play a special part in the evolution of humans (see chapter 6).

All life forms, from fruit flies through amphibians and mammals, have bits of viral DNA spliced into their cellular DNA. These virus remnants are left from retroviral epidemics that occurred eons ago. Even our own genome contains countless pieces of DNA that do not code for blue eyes or blond hair but for a dormant retrovirus. Such DNA is called "proviral" (or proto-viral) because each piece, if big enough, can become an active virus under certain circumstances. Even after thousands of years, it can emerge as a virus particle, once the proviral DNA has been set free from the cellular DNA. Virologists can point to cases of such emergence, although apparently it is a rare occurrence. In such cases, we say that an endogenous virus (one born inside the host) becomes an exogenous, or circulating, virus. An insider becomes an outsider, and the emergent virus may or may not be harmful.

Viral genes spliced into the host's cell genome may protect its host and may have even created mammals. We know, for example, that endogenous viruses protect certain mice against attacks from exogenous retroviruses of the same family. The insider protects its host against a related outsider much as vaccine virus like cowpox protects against the more aggressive smallpox virus. We know also that a certain retrovirus causes humans to experience the taste of some starches as sweet.

By far the most important retroviral contribution to the living world is the mammalian placenta. The placenta is kept intact by retrovirus proteins. This amazing structure allows nutrients to enter and waste products to depart while at the same time protecting the fetus against the entry of many substances that would cause harm (see chapter 6). Without the placenta, mammals would not be mammals.

Just as phages may help bacteria, so retroviruses may help a host to better adapt to a changing environment. The two fitness champions do not help out for charitable reasons, of course. By making themselves indispensable to their host, they guarantee their own survival as well.

Viral genes may define the host's ability to adapt. Retroviruses also seem to play a role when it comes to the rate at which a living being loses DNA. The length of a living being's DNA is not proportional to the complexity of the organism, the number of its genes, or its rank on the evolutionary ladder. An amoeba has a DNA molecule with a size worthy of a whale. The length of a DNA molecule depends on the size of the so-called junk DNA, rather than on the number of genes which the particular organism needs. Most of this junk DNA codes for retrotransposons, which are loose pieces of DNA that move around like retroviruses inside the cell DNA. The length of a species' DNA seems to a large extent dependent on the rate at which a specific species loses its junk DNA, in particular, its retrotransposons. These are actually primitive retroviruses, which perhaps make it possible for a species to quickly adapt to changes in the environment.

Viruses may have served as models for the way in which life on earth has developed. It is generally assumed that life coded by DNA, as we know it today, originated from so-called ribo-organisms that used RNA as code and RNA as enzyme as well (a function which is handled by protein in "our" world). The life of bacteria is based on a circle of double-string DNA. The code of a living cell lies anchored in a linear double-string DNA molecule. Retroviruses get their name from an enzyme (reverse transcriptase), which is able to transcribe single-string RNA into linear double-string DNA. This proviral DNA not only integrates into the cell DNA, but double-string DNA circles are also formed in the cell core. It would seem obvious that a retrovirus enzyme, and maybe even a primitive retrovirus, must have existed before any of the life forms coded by DNA, that is, if we believe that our DNA world originated from an RNA world. If that is true, retroviruses can rightfully lay claim to the title of "the fittest of all."

Viruses do not always take such good care of their host. Whether a virus makes you sick or not depends on the efficiency of the spread of a virus in its host population. If a virus is not very infectious, the host has to be exposed to large numbers of virus in order to get infected. The host may not always be able to cope with so much virus. The AIDS virus is the most dramatic example of this sad fact. Yet most retroviruses are innocent; even most viruses of the HIV family are innocent. A majority of the lions in the Serengeti park in Africa carry an HIV-like virus all their lives without suffering a bit of pain. The AIDS virus, on the other hand, shortens the average life span of a child born around the year 2000 in the southern part of Africa from 65 years to about 40 years. Between 1990 and 1999, mortality among young people between the ages of 25 and 35 in South Africa increased at least sixfold due to AIDS, with dramatic economic and social consequences for the entire population.

Viruses surf the living world to survive. All misery of AIDS is the result of an innocent simian virus that happened to jump to human beings. It was apparently forced to make this move by the fact that its natural host, the chimpanzee, is becoming extinct. It had fewer chimpanzees to infect, and human encroachment on jungle habitats made us a handy alternative. Humans are the reason that chimpanzees have come close to extinction, by eating them and by cutting down the rain forests. Like our fellow primates, we have on some of our cells a receptor that the simian virus can use for entry. It is not there to allow infection, of course; but whatever its original purpose, it now admits HIV. And in humans, the cells with this receptor are immune cells: the cells that fight all kinds of infection. So HIV is more deadly for us than are most viruses—and more deadly, by far, than it was in its simian hosts.

Genetics are the natural but slow way for the host to protect against viruses. A few lucky humans lack the receptor that allows HIV to enter human cells. Descended from a group of inhabitants of Northern Europe, these people are virtually impervious to being infected with the AIDS virus. The AIDS virus will therefore never eradicate all of mankind, even if we never find a way to stop it. In fact, the peak of the epidemic has passed already in some parts of the world, specifically in Northern Europe, where natural resistance to the virus occurs most frequently. In Africa no people live that have a genetic resistance against HIV. On that poorest of all continents the AIDS epidemic affects both the lives of the infected and the uninfected and therefore threatens the future of the African people.

New viruses will come because the human population keeps growing. In the years ahead, AIDS is not our only worry, as we can see from much smaller epidemics: Ebola virus, West Nile virus, SARS, and Monkeypox. Many other exotic viruses have appeared and many more must be expected. The root cause is the unstoppable growth of the human population. This growth goes hand in hand with the destruction of practically all areas where animals can live in the wild. If the viruses of these animals get the chance—which we ourselves provide by pushing into animal habitats—they will jump to us.

Many will fail to make the jump successfully, but a few will not. If just one can make the jump and adapt to us as its new host, its descendants—which inherit its adaptability to humans—will soon number in the millions. Our danger is enhanced by the increased density of human populations, a natural side-effect of growth. The closer people live together, the more easily a virus spreads among them. But some viruses give us more options than others.

Viral fitness and our way of life. As described in the first five chapters, we are exposed to many viruses just because of what we need to survive—air, food, and water—and because of weather conditions, over which we have little control. Such viruses give us few options. The viruses of chapter 6—the simple retroviruses—give us no options at all, but luckily they never harm humans.

In contrast, the viruses of the last four chapters give us lots of options. We can avoid their dangers, to a great degree, because they are linked with largely optional sexual behaviors, territorial aggression (against animals and against fellow humans) and our desire for exotic foods.

The two big questions are: can humans gain the insight and the will to avoid the avoidable viruses?

And can we also arm ourselves against the viruses that will keep coming, as escapees from the wild or as weapons of man?

I

AS NATURAL AS BREATHING

The Flu Virus

The virus that causes our annual epidemics of influenza comes to us through the air. We think of the flu as a human disease, but its agent is by nature a virus of birds. Circumstances may force this avian virus to infect other animals, and it gets better and better at doing this. But it is most at home in wild ducks. Ducks are infected very young, and about 20 percent of any wild flock usually harbors this avian virus. It is endemic in ducks—a permanent feature of their world—but never makes them sick.

While our flu virus infects cells of the airways, the avian flu virus infects cells of the stomach and intestines. It enters birds when they drink water and is excreted with their feces. In lakes where ducks get ready for their annual migration, influenza virus is present in large quantities, and the little beaches around the lakes are full of very infectious excrement. However, once the birds migrate to warmer regions, their flu virus goes dormant. Preferring cooler weather, the virus dwindles to small quantities in the wintering ducks—but it breeds up again when they return to their home lakes.

Contact with wild birds or their excrement causes epidemics of flu virus infections with some regularity in seals, whales, horses, chickens, turkeys, and pigs. In contrast to what happens in a population of ducks, where the virus is constantly present, in these other animals the virus is generally absent until it suddenly emerges. The same is true for people.

Flu epidemics in pigs are of most concern to us for several reasons. The flu virus cannot thrive long in a group of seals, whales, or other mammals that have very limited contact with each other. Its transmission is too dependent on intimate contact; pigs like living close together, so they sustain more flu epidemics. Also, pigs are a major food animal and in some countries, such as China, they are by far the main animal raised for meat. Compared with most other food animals, such as chickens and turkeys, pigs are large and expensive—so one stricken pig is a greater economic loss than quite a number of stricken fowl.

Our biggest source of concern is the fact that hog viruses can more easily infect people than the viruses of most other animals. Workers in slaughter houses regularly contract swine flu infections. Children are susceptible to viruses that seem to be a mixed form of bird, hog, and human viruses. And we must not only fear the true hog viruses but also those of other animals that happen to infect hogs—including the bird flu viruses.

Apparently transmission of the influenza virus from bird to pig makes it easier for the virus to take the trip to man. The flu viruses that annually plague us seem to arise in China where both ducks and pigs often live closely with small farmers. As in other very poor farming cultures, animals may even live in the house, especially when the weather gets too cold to keep them safely outside. One virus that arose this way caused the 1918 flu epidemic, though its history was not known for years. And while pigs make a good stepping stone from birds to humans, bird viruses can sometimes infect humans directly.

How does the versatile flu virus do so much damage? Although basically a bird virus, its disease strikes humans hard every winter, causing a worldwide epidemic—or pandemic—that varies in severity. It usually strikes in winter because in most places that season brings the conditions it likes: relatively cooler air with a low degree of humidity. The virus enters the human body through the airways as we breathe. Once inside, it immediately finds cells in which it can multiply: the cells of the epithelium or mucosal lining. Fever and inflammation result, and the irritated airways cause lots of coughing and sneezing. Only one out of every ten flu virus infections takes its course without us noticing that anything is amiss.

Flu becomes most serious when the virus attacks the smallest branches of the airways. It is most virulent in these cells located deepest in the lungs. In this area, it takes only a few virus particles to have a big effect, since the cells have the highest density of receptors for the flu virus. The virus multiplies at a furious pace in a couple of days, by which time the mucus in the nose contains tens of millions of virus particles.

Transmission of the human flu virus takes place during the first week of infection. Flu patients are most contagious during the first three or four days of the illness. After a week, there are no more virus particles to be found, but by that time, they have been sneezed and coughed all over everyone nearby. They have been spread through the air by aerosol particles of mucosal fluid. This method of transmission makes the human flu virus more contagious than the duck virus, which is spread by water.

The clinical aspect of the flu—the eruptions of coughing and sneezing—thus contributes directly to the spread of the virus. And flu is an illness against which you cannot easily protect yourself by adapting your behavior. If you do not want to contract HIV infection, you can avoid unprotected sex or the use of injectable drugs, but the only way to avoid flu is to give up breathing. You could stay away from other people, but this is not easy.

Suppose you board a bus or train and sit down next to a perfectly healthy-looking person. He is not covered with spots, for example, as he would be with measles or smallpox. After awhile, he begins to cough and does not cover his mouth. Coughing does not prove he has flu but, even if it did, are you likely to change your seat? If you do, and the flu is going around, will your next seat companion be any healthier than the first one?

No other mammal is as mobile as man. For almost any virus, man is the ideal vehicle for getting disseminated all over the globe. Ducks can only take the flu virus between their two seasonal homes, but people can take it anywhere. And every person, of whatever age, sex or race, can get the flu. Adults spread it mainly by traveling, especially in airplanes. It is hard to imagine a better way of spreading the flu than by airplane travel. It can take an influenza virus to any corner of the earth within twenty-four hours.

In 1977, an airplane full of passengers had to operate without ventilation for about four hours. In that short period of time, 72 percent of the passengers got the flu from one fellow passenger who was coughing a lot. The virus cannot have spread in any way other than through the air, since the passengers hardly left their seats. And while that particular plane had a ventilation breakdown, many planes use only minimal ventilation systems all the time.

However, the first infections in a flu outbreak do not usually take place on a plane. They take place among children in school, who take them home to brothers and sisters at home, to parents and grand-parents. By way of the adults, the virus then spreads from one location to another. Newborns and the elderly tend to be last infected, and the elderly in particular lose large numbers of their population to the flu.

Studies in Houston, Texas, have shown that during the average flu epidemic in the western world, 40 percent of young children and about 20 percent of the adults become infected. At least 90 percent of infected people have symptoms, but most never require medical attention. Nevertheless, some 20,000 people in the United States die from the flu every year. The rate of infection and morbidity, or disease, is highest for young children; but the rate of flu mortality is highest among the elderly, especially those with heart problems or cancer. Pregnant women too are at higher risk of flu death than the general adult population.

The epidemic of 1918 departed from the average in major ways. After three quarters of a century, it remains the worst ever to occur, causing death in 30 to 50 million people worldwide. The most conspicuous thing about the 1918 pandemic was that it particularly decimated people in the age group of 20 to 40 years—people who had just died by the millions in The Great War. Ultimately the epidemic claimed more victims in a short time than any other infection in the history of the world. Yet unlike plague, the flu was not usually a killer; it had always been an illness cured by staying in bed for a few days. People think of flu as simply a bad cold; indeed, in Italy it was originally called *influenza di freddo* (cold influenza). But 25 percent of the entire American and European populations contracted the flu in 1918, and one in every fifty died from its consequences.

The 1918 flu virus was far more virulent than any other before or since. The virulence of a virus has to do with the damage it can do: its capacity to cause illness and death. One school of thought holds that viruses do not benefit from being too virulent—from causing too much harm to their hosts, on whom they depend for their reproduction. This school of thought assumes that a virus and host eventually adapt to each other so that the initial, or acute, infection takes its course almost or entirely unnoticed. This theory is only partly correct. With the flu, the virus spreads strictly during the first week of infection, then begins its cycle in a new host. Thus the survival of this virus is not compromised by its virulence. Once the virus has moved on, it has no problem if the lungs in the old host are reduced to an amorphous foaming mass, making it impossible to draw even the slightest breath.

In 1918, there were actually two influenza epidemics. The first occurred in the spring and was in no way different from a normal annual flu epidemic. The second occurred in the fall, with the tragic outcome that millions died all over the world. This disaster would become known as the Spanish flu, because the first few cases were reported from Spain. But as we will see, each year's flu virus actually arises in China and causes at least a small epidemic there before spreading abroad.

Largely unnoticed in the 1918 chaos was the fact that pigs also had a flu epidemic about that time. And ever since then, a swine epidemic of influenza has occurred nearly every year. Did we catch the flu from the pigs or did they catch it from us?

The latter now seems to be the case, though that discovery was years away in 1918. First the flu virus had to be discovered, which happened in the 1930s. Virological techniques had to be developed, so it was 1936 before researchers could replicate influenza viruses in the laboratory. To obtain sufficient numbers for systematic study, scientists would inject saliva containing the virus into the amniotic fluid which surrounds a chicken embryo. When breathing, the embryo takes in and expels this fluid. The virus thus grows in its lungs, although in nature a flu virus infects gastrointestinal cells in birds. Within two days, it would be possible to see if there was virus growing: the amniotic fluid would become non-transparent and opaque.

In 1944, the first Americans were vaccinated against the flu. The vaccine was based on that year's flu virus, which had been cultivated in chicken eggs, then "killed," or de-activated. Like all vaccines based on killed virus, it contained enough viral substance to stimulate an antibody response but could not cause disease, because a killed virus cannot reproduce.

Since 1944, a routine has developed. Each year flu specialists watch for signs of the annual epidemic and its cause: the virus we need to fight. Then they race against time to cook up a vaccine based on that virus, hoping to be ready before the epidemic gets out of hand. In about 8 months, the vaccine has to be produced in hundreds of millions of dosages.

What made the 1918 flu virus so virulent is still mysterious, but continuing research has brought progress in our understanding. The answer remains important because at any time, the annual flu virus could turn out to be a killer of 1918 proportions. It would not be the same virus as in 1918, but the more we know about that virus, the better we can manage each new one.

An early question for flu researchers was why that virus killed people 20 to 40 rather than older people. A simple answer was that people older than 40 had already survived an earlier infection with a similar virus. They were thus somewhat immune to it. This suggests that a 1918-like virus was present earlier and went unnoticed. Indeed, a search of records found that small flu epidemics had occurred in 1916 and 1917 in British army barracks in England and France.

An additional answer relates to a lethal sleeping sickness, *encephalitis lethargica,* which was described in 1917 by Baron Constantin von Economo. According to this Viennese physician, patients with the disease would sit motionless and mute. In 1982, Ravenholt and Foege, two scientists at the U. S. Centers for Disease Control (CDC) showed that living through a serious flu was directly associated with contracting encephalitis lethargica.

As often happens in science, they had found new implications in an old story. On November 7, 1918, a steamship from New Zealand landed in Western Samoa. Its crew brought influenza, and 8,000 West Samoans died from it in two months. When the nearby inhabitants of American Samoa heard about this, they sealed their island from the outside world and kept the flu out. During the 1919 to 1922 period, 79 Western Samoans died from von Economo's disease, but only 2 died in American Samoa.

The virus which caused the 1918 disaster was neuro-virulent: it had a preference for infecting brain cells *in addition to airway cells.* This characteristic could hold the key to predicting whether next year's flu virus might be the next 1918–like killer.

Getting back to the simple answer: if people over 40 were relatively immune to the 1918 flu, there must have been flu viruses circulating decades before that. Since all flu epidemics since 1918 have originated in China and, more specifically, in the southern Chinese province of Guangdong (formerly Canton), researchers looked to that province for earlier epidemics. Sure enough, the first flu epidemic ever described in detail took place in September 1888 in Guangdong. It is likely that there were even earlier ones, never recorded.

In any case, when influenza struck Guandong in 1918, it was no worse than any annual flu in the area. Moreover, it struck teenagers more than older people. The first finding suggests that people had earlier been struck by such a virus, so the population had some immunity. The second finding suggests that immunity was strongest in that part of the population that was born in the second half of the 19th century.

How did the 1918 flu virus travel the world in those days before airplane travel? Nothing is documented, but Cantonese workers came by boat to France to dig trenches for the war, so perhaps one of them brought the flu. If so, the first European cases occurred in France, unnoticed, and were reported from a slightly later outbreak in Spain. Perhaps the flu then spread to North America with service men returning after the war.

Researchers began to suspect that the first truly human flu viruses originated early in the 19th century but did not become a threat until the early 20th century. What animal was its host before it became a human virus? The first step toward an answer was to identify the 1918 virus, using a phenomenon known today as *original antigenic sin.* Christians believe humans are marked by the "original sin" in the Garden of Eden. By analogy, we are all antigenically marked by our first flu infection. This usually occurs in childhood, and every subsequently occurring flu virus infection peps up the antibody response against that first virus.

So, decades after the 1918 epidemic, specific-antibody tests were performed on elderly people born at that time to determine what kind of virus had infected them in those years. It turned out to be what we now call H1N1 virus.

Flu viruses are viruses with eight RNA strings coding for ten proteins. Two proteins cover the virus surface in a mosaic distribution. The rest rattle around loose, inside the virus. The surface genes, H and N, determine the basic nature of the flu virus and are always represented in its name. The H gene codes for hemagglutinin, and the N gene encodes neuraminidase.

Hemagglutinin is the crucial protein that enables the virus to bind to the cell surface. It thus determines the species specificity of a particular flu virus: which cell in which animal the virus will infect. This is because binding, entry and infection all depend on there being a cell receptor that hemagglutinin happens to fit, like a key in a lock. The receptor is not there to admit a virus; it has some beneficial purpose for the host—but the virus happens to be able to use it. Cell receptors are like the cat-door that lets your pet in and out of the house: they can also let in a much less welcome creature.

Neuraminidase is an enzyme, as indicated by its -ase ending. Like all enzymes, its job it is to eat things. That is, it digests or dissolves certain substances, and this can have a constructive or destructive effect. In the flu virus, it is constructive for the virus but destructive for the host cell. As new virus particles are produced inside a cell, they are chemically attracted to the cell wall. They line up there, nudging against the wall, and their neuraminidase munches right through it. This destroys the cell and releases the virus particles to spread in the body.

After H1N1 of 1918, there was H2N1, H2N2, and so on. The genes inside the flu virus make proteins crucial to its virulence. Some act directly, by causing a symptom; some act indirectly, by impeding the host's defense mechanisms. Because the genes inside the flu virus are loose, they can exchange bits within one gene, called recombination, or they can rearrange the eight RNA strings of the flu virus. For example, a hog virus can provide a gene, a chicken or a goose virus another gene, and a human virus a couple of others. This is why new flu viruses keep coming, always with new characteristics. Whatever the combination is in a given year will determine the virulence of the virus, how easily it is transmitted within a species; how easily it attaches itself to a host cell, and how easily it avoids being recognized by the immune system.

To trace the history of H1N1, its genetic code had to be deciphered. The necessary techniques were not available until the last two decades. Then RNA had to be found from an actual flu virus that had infected someone in 1918. Success eventually came to Jeffrey Taubenberger of the Armed Forces Institute of Pathology in Washington, D.C. In the archives of this institute, material was stored from 74 victims of the 1918 flu epidemic. Of these, two samples were suitable for study and found to be positive for influenza RNA. Meanwhile, Johan Hultin obtained tissue from a woman who had been buried in a mass grave of 1918 flu victims in Alaska. This woman's lungs were still in very good condition some eighty years later, because she had been buried in permafrost.

Study of these three persons confirmed the H1 identity of the 1918 hemagglutinin gene. This H gene clearly belonged to the group of genes from viruses that have caused epidemics in people over the decades since 1918. However, the one virus whose H gene resembled it most was a hog virus isolated in 1930. This confirms the view that, after humans contracted the 1918 flu virus, we subsequently infected pigs.

The genealogical tree of flu viruses has two main branches, one with all the bird viruses and one with swine and human viruses. If avian flu viruses, and thus birds, were the source of the flu viruses now adapted to humans and pigs, the genealogical tree showed that the avian precursor of the 1918 virus infected its first human about 1905. We now know that of all the flu viruses ever found in people, the 1918 virus is the most like a bird virus.

Researchers had long assumed that the exchanging and re-arranging of the eight loose RNA fragments in the virus particle determined the character of each flu virus. Mark and Adrian Gibbs, a father and son team from Australia, very recently came up with a revolutionary idea for explaining the aggressiveness of the 1918 virus. They envision that, in addition to the exchange of whole genes, there is formation of entirely new genes; this is done by pasting parts of a gene together rather than swapping an entire gene, and it can happen not only with the internal flu genes but with H and N on the surface.

There is really only one way in which all this could take place: recombination. This happens when a host is invaded by two virus populations—for example, avian flu virus and human flu virus. One virus from each population penetrates the same cell, which is possible if they arrive at the receptor at the same instant. Simultaneous infection sounds improbable but is not so unlikely, with millions of viruses jostling for millions of cells.

Once inside the cell, the two different viruses simultaneously reproduce, exchanging bits of genes, and ultimately there is a new virus. It may be a weakling; it may be a dud that does not even survive; but it may be a super flu virus, fitter and perhaps more lethal to people than its forebears.

The Gibbs team suggests that in the 1918 virus, a part of the H gene was from a swine flu virus and another part from a human virus. Assuming that a bird virus managed to infect both people and pigs in about 1905, the result would have been an H1N1 hog virus and an H1N1 human virus. Since a similar flu appeared in army barracks in 1916–17, someone already infected with the human virus must have become infected with the hog virus, combining the two viruses and creating the 1918 virus.

The human-swine flu combination somehow formed the basis of the lethal character of the 1918 virus. It also made the virus better able to spread than any other flu virus. It was no longer limited to one species. Thus, while causing the 1918 epidemic, it also infected pigs and caused a swine flu epidemic. At some point it entered a pig already infected with a true hog flu virus, and from this marriage came the H1N1/Iowa 1930 virus. This virus, isolated from an epidemic in Iowa, has long been known to resemble closely the 1918 human flu virus. It still thrives, having crowded out all other hog flu viruses. The human virus that caused the 1918 epidemic has not thrived. It became extinct in man after the 1918 epidemic.

The Gibbs scenario is not implausible. Since 1918, history has brought us more and more evidence for the exchange of influenza viruses among birds, pigs and humans. It seems that bird viruses end up in people, via pigs, each time there is a serious pandemic. This happened in both 1957 and 1968. The explanation lies not only in the genetic composition of these flu viruses but in the receptors on the cells in these three species. Although the 1957 and 1968 flu viruses were in part bird flu viruses, they were not entirely bird viruses—and that was the secret of their success. If they had been bird viruses, they could not have entered human cells. Our cells lack the NeuAc-2,3Gal receptor, which is essential for allowing our infection by avian flu viruses.

Swine, on the other hand, have this receptor for bird flu viruses. Not only that, they have the 2,6Gal receptor for human flu viruses. So pigs can become infected with both avian and human flu viruses. And we know that hog flu viruses can infect people. In 1976, this became painfully obvious once more, when it was feared that an epidemic due to a swine virus would erupt among military recruits in America. On the basis of this fear, 40 million Americans and a considerably smaller number of Europeans, specifically in The Netherlands, were inoculated against hog flu. But it was a false alarm: no epidemic took place.

True hog flu viruses, as well as avian and human viruses, continue to exist among pigs in various countries, specifically in Europe and Asia. In the 1970s, H3N2 viruses were transmitted from humans to pigs in Europe. Over the next two decades, a bird virus of the H1N1 type was endemic among pigs in Europe. So nobody was much surprised when, in the mid-1980s, pigs in Italy were found to have viruses which looked like human H3N2 on the outside and like bird or hog H1N1 on the inside. These avian-human reassortant viruses caused flu on a limited scale among Europeans during the next few years. They spread among pigs in China and emerged as the flu virus of a child in Hong Kong, who was infected in 1999.

The story becomes even more complicated. Chinese pigs were found to be infected in large numbers with this H3N2 virus and also very often with a second virus, H9N2, which had originated in birds, probably ducks. The latter virus was widespread among all kinds of fowl in the markets of South China, specifically in Hong Kong. It also was able to cause flu in people, as was found in 1998 in Guangdong and in 1999 in

Hong Kong. Fortunately, none of these H9N2 infections was serious and all victims recovered quickly.

In 1997, however, a virus had emerged in Hong Kong that looked much more ominous than any other influenza virus. It was an avian virus, H5N1, which not only infected birds but made them very sick. Of all the chickens infected, 70 to 100 percent died.

It also made people sick and of 18 infected, 6 died. The virus seemed to be transmitted directly from chickens to people. This was very bad news, but luckily it could not spread person to person; each human case resulted from exposure to a chicken. The disease struck persons between 1 and 60 years of age. It seemed to bypass the elderly, perhaps because they seldom bought fowl in the markets. But among adults up to 60, the risk of death increased with age. The first person to die from the H5N1 bird virus was a 3-year-old boy, but other deaths were a girl of 13, a young woman of 25, a man of 54, and a man of 60. Further disaster could be prevented only by slaughtering millions of chickens at thousands of poultry farms and markets in the Hong Kong vicinity.

The 1997 virus had a short run, but it seemed to be even more dangerous to humans than the 1918 virus. Was it now possible for some avian viruses to infect humans directly, without first making a stop in pigs? And if so, were these bird flu viruses more lethal than any other flu virus seen by man?

Some 30 years before that, the 1968 epidemic of Hong Kong flu had featured an H3N2 with a totally new H gene. The epidemic of 1957 was caused by an H2N2 virus with a new H gene and a new N gene. In both cases, as in 1918, avian viruses were able to infect humans after gaining genes or parts of genes from flu viruses of pigs or humans. But the Hong Kong virus of 1997 seemed to be entirely a bird virus.

A new phase seems to have begun in the evolution of avian flu viruses. They have found their way directly to man. The H5N1 of 1997 that had killed chickens in great numbers around Hong Kong was identical to the virus that infected 18 people—except for one amino acid. A change in this one bit of one protein, near the spot where viral hemagglutinin attaches to the host cell, was apparently enough to make a true bird virus spread easily to people. If the virus could not spread *among* people, that was probably because it needed direct contact; it was not sufficiently infectious when spread by aerosol particles.

How do we know the H5N1 virus could not spread person to person? Because each of the 18 persons was infected by a distinct strain of H5N1.

So several infections from different sources had occurred. Direct contact between chickens and people was needed. The final proof of this came when all chickens in Hong Kong were slaughtered. Immediately there were no more H5N1 infections in people. The danger of an epidemic had passed, despite the fact that not one human being was immune to this virus.

The great aggressiveness of the H5N1 virus was based on its H gene, on the one hand, and on an internal gene, on the other. As early as the 1970s, this virus was known in Chinese ducks and geese. The H5 gene seems to have originated with the geese. The source of the N gene is still unknown, but the eight internal proteins seem to have come from a quail virus of the H9N2 type.

The H5 gene has been stable over the years in all its avian hosts, which suggests that it is very well adapted to geese and even to chickens. However, the internal genes of H5N1 viruses are highly subject to change, which suggests that they are still adapting to chickens. Wild birds, such as ducks and geese, seem to be the true pools in which this and other avian flu viruses thrive. Domestic animals such as chickens only function as a leg up to man, as is the case with pigs. Ordinarily, a duck or goose virus cannot be transmitted to man, but a hog or chicken virus can. The flu virus spreads death and destruction among these new intermediate hosts but still causes no disease in wild ducks and geese.

Does the flu virus derive some benefit from making a trip, on a somewhat regular basis, from wild water fowl to domestic animals and from those animals to man? Or is the trip only an accidental event without any meaning for the survival of avian flu viruses in general?

We do not know. But we do know that the flu viruses of ducks and geese evolve very slowly. The influenza virus feels at home with these water fowl which live in the wild, and it no longer needs to adapt to them. After a trip to chickens, horses or people, could these viruses return to ducks and geese as new strains? Might such strains even make them sick? Do duck and geese flu strains ever become extinct? All these things seem unlikely because their well-adapted viruses are so stable.

Meanwhile, new flu virus families have formed, based on avian or part-avian ancestors. They have settled permanently in pigs, horses and other mammals since the beginning of the 20th century. The diversity in species of flu viruses has increased, because the number of infected species has increased. At this point we could never wipe influenza viruses off the face of the earth. They are able to spread rapidly, they can move across the barriers between species, and they no longer have just one or two pools (e.g., ducks and geese).

The flu virus has ensured eternal life for itself as a virus by jumping from one host to another. More and more flu variants have apparently learned how to do this trick, and not only in China. In 2002–2003, chicken farms in the Netherlands were struck by the avian flu strain H7N7, and soon more than 100 humans had been infected. The chicken virus H7N7 had the unique and dangerous ability to jump from human to human. A Dutch veterinarian died and, by ominous coincidence, H5N1 claimed one more Chinese victim in the same year.

The H7N7 infection soon showed up at some Dutch pig farms. In the Netherlands, there are over 10 million pigs and even more chickens; some farms raise both types of animals. As for humans, the country is one of the most densely populated in the world. All ingredients are in place for the next killer flu virus, and this time it may not be from China.

2

FARMING AND FEEDING THE HUNGRY

Plant Viruses and Human Enteroviruses

It is generally accepted that agriculture arose between 10,000 and 7500 B.C. in Mesopotamia, the "fertile crescent" between the Tigris and Euphrates rivers. In fact, the region's name, in Greek, means "between the rivers." Now it is largely occupied by Iraq, but some Biblical scholars think it was the original Garden of Eden.

When humans moved from hunting and gathering to farming, it was a momentous step. Many factors contributed, but climate change was no doubt among them. Perhaps more people were able to reach adulthood, due to the milder conditions which began after the last ice age, so hunters increased in numbers and decreased the number of prey animals. The men had to go farther and farther to obtain meat, and crops could provide more food near at hand.

Whatever the cause, people first tamed grasses to provide grain. Above all they needed grain to make bread, though some archaeologists say they wanted it even more to make beer. In those ancient times, the poor often lived on little more than bread and beer, as we know from records on Egyptian pyramid workers.

As a result, fields were cleared and burned to be used for crops. There were so few mouths to feed in those days that it was possible for this primitive form of agriculture to endure for a long time. These and other techniques spread within a couple of thousand years to Europe, Africa and Asia, or perhaps were discovered independently in these areas. Agriculture surely arose independently in the Americas, and somewhat later. It was apparently not influenced by Asian agriculture, because although the New World people migrated from Asia, they did so as hunter-gatherers.

Around 3500 B.C., towns and villages began to form along the Euphrates, Tigris, Nile and Indus rivers. Some became big cities and trading centers. From then on, ever larger groups of people and livestock were sustained by agriculture that was practiced on an ever-larger scale. However, the greatest revolution in agriculture took place two millennia after that, when Europeans discovered the New World.

The Age of Exploration brought a massive flood of new plants and crops to Europe, Asia, and Africa: corn, potatoes, cassava, all kinds of beans, tomatoes, peanuts, pineapple, and cacao. Other crops found their way from the east to the west: rice, wheat, oats, and barley. In addition to a wider selection of food stocks, there was a massive increase in the number of births around the world. This increase was the result of improved nutrition that brought earlier sexual maturity and a longer period of fertility.

During the period from 8000 B.C. to the beginning of the Christian era, the world's population had increased steadily from some 5 million to 150 million. In the year A.D. 1000, world population stood at one quarter billion, and this number had doubled by the year 1600. Between 1600 and 1800, the population of our planet increased by another quarter billion. In 1800, the population had not yet reached 1 billion, but in 1900, it was just a little over 1.5 billion. By 1950, it had gained another billion, and 2000, it stood at 6 billion. Now we are facing an unprecedented increase in our population, which is projected to reach 8.5 billion by 2100.

Did the development of agriculture and husbandry make possible such population growth, or was it population growth that led to these developments? More mouths to feed means more food to supply, and one thing is sure: this is not possible without agriculture and livestock. In his book *How Many People Can the World Support?* Joel Cohen expresses amazement at how humans learned to domesticate plants and animals—to breed sheep, goats, cows, pigs, and chickens to meet our needs. He is likewise amazed at how these animals may have used us for their purposes. After all, people have cultivated such a variety of grains and tubers to sustain those animals, which would otherwise have had to forage for themselves.

Whichever came first, agriculture or population growth, there are now more people alive on this earth than the total of all who have died since the beginning of humanity. This has enormous consequences for the food supply, and the question is how long such population growth can be sustained. At this point we must wonder how agriculture can keep up, how the world's people can have enough to eat. Already, many do not have enough to eat or have such limited diets that they are seriously undernourished.

Two groups of viruses complicate our food problems. First, there are plant viruses and viroids that can destroy harvests and cause famine. Second, there are human enteroviruses that are linked with malnutrition or contaminated food. To a great extent, we cannot escape from these viruses because we must eat, although we can minimize the problems they cause. With the plant viruses, we can control our increasing drive toward monoculture. With the human viruses, spread mainly through feces, we can improve public and private hygiene to lower their risk. But both efforts are less and less likely to succeed as more and more of the world's people live in poverty.

It is no surprise that viruses infect plants as well as people and other animals. Like all living things, plants are made of cells, and viruses need cells in which to reproduce. Plant viruses are often harmless to plants, but they can see to it that there is little or nothing to eat over a wide area. Their survival at the expense of all kinds of crops can cause famine or, at the very least, a major economic loss for farmers. The annual loss worldwide of tomatoes, peanuts, tobacco and many other crops as a result of infections with the tomato spotted wilt virus is estimated to be at least one billion dollars. Other viruses regularly cause the failure of crops such as sugar cane, corn, cassava, cacao, grains, tomatoes, beans, plums, apricots, and peaches.

Plant viruses basically look and act just like viruses of humans and animals. And like those viruses, they occur worldwide but seem to afflict Africa most severely. This is mainly because, though sometimes they spread in plant seeds, plant viruses spread mostly by insects—and insects thrive in warm and humid places like Africa. Plant viruses do not cause illness in insects—they do not even infect insects, which simply transport the viruses from plant to plant. Nor do they infect any animals, except for nanovirus, a very small plant virus whose mutations have allowed it to replicate in animals under the name of circovirus.

Survival of the worldwide human population is minimally affected by plant viruses because most famines have only a local effect. This is not to minimize the suffering of the thousands of people who can die in such a disaster, and in one historic case, a plant pathogen had worldwide effects. It caused not only a vast famine but a vast migration of people. It caused the potato blight and the Irish Potato Famine of 1845–1852. Actually, though viruses had paved the way, the blight was not caused by a virus but by a fungus, *Phytophthora infestans*, that originated in the Mexican Highlands.

Moreover, the blight did not start in Ireland but in the United States, in 1843, on a small scale. This was not known to the provincial government of Flanders, which that year imported new potato strains from America. In June 1845, the potato blight was observed for the first time in Belgium. In a month or two, the disease had blown over to The Netherlands and France. By August, it had gotten to England, and toward the end of August and in early September it hit Ireland.

Potato blight can destroy a potato field in a few days. As of 1845 and for the next few years, Ireland did not have enough food because the bulk of the potato harvest had turned into black slime. The wealthier Irish could switch to corn and other foodstuffs, but the Irish poor suffered illness and starvation from lack of food. They also suffered from diarrhea and cramps when desperation forced them to eat rotting potatoes rather than nothing at all.

Why was the blight especially bad in Ireland? The cause of this famine was partly a result of the population growth in those days. But rather than the growth itself, the cause was the increase in population density. The number of mouths to feed in Ireland simply outstripped the ability of the Irish soil to feed them.

Also, at that time, potatoes in Ireland were the staple food. Not only did everybody eat them every day, but potatoes served as feed for pigs, horses, cows and chickens. Around 1840, 40 percent of the Irish population, some 3 million people, was totally dependent on potatoes for their food supply. Men who did hard physical work often ate four to five pounds of potatoes per day, and sometimes they were paid in potatoes. In 1845, 35 percent of all agricultural soil was used for the cultivation of potatoes. In 1847, after the great potato blight, this had gone down to just one percent.

Another factor was that amidst the famine, the undernourished Irish population was afflicted by one of the severest winters in European history. They also suffered from cholera epidemics, spread by diarrhea, and infections with borrelia and rickettsia that were transmitted by the lice that plagued poor and crowded hovels. Just during the winter of 1846–1847, the death toll among the Irish people exceeded five hundred thousand.

But the root cause of the famine was not the fungus but the sharply decreased diversity of potato strains that Irish farmers could use in the middle of the 19th century. The potato had come long before that from the New World, where indigenous peoples raised hundreds of varieties. By the early 19th century, Irish farmers themselves had developed as many as two hundred potato variants. Such diversity would have protected them against any blight, but a series of viruses and viroids attacked one variant after another. Eventually, the Irish relied on one single type of potato that resisted these attacks. Called the *lumper*, it was not much of a potato. It was watery and not very nutritious; even pigs scorned it unless they had no alternative. Worst of all, though it had resisted the viruses and viroids, it turned out to be extremely sensitive to potato blight.

The disastrous concurrence of monoculture, a watery potato, the potato blight, the epidemics, the severe winter, and the great Irish population density caused a famine which was beyond the control of any government, certainly the English government of that time. During the 1847–1855 period, millions of Irish left their home country, mainly for America.

By the end of the century, the diversity of potatoes had seriously dwindled even in the United States, again due to viruses and viroids. American farmers began to look for new potato strains which would be more resistant against the potato blight. Like earlier farmers, they went to Mexico, the cradle of the potato. They returned from Mexico with potatoes resistant to *Phytophthora infestans*, but these turned out to be infected with the *potato spindle tuber* viroid. Although by that time the potato blight was apparently under control, this dangerous micro-organism can cause disease in potatoes just as serious as potato blight. It also infects tomatoes, another product of the New World.

How are viroids different from viruses? They are the smallest infectious particles in existence, as far as we know. As indicated by their name, they are "virus-like." They consist of very small circles of RNA which, unlike viruses, do not code for any proteins and are enclosed in no outer shell. At first we thought they infected only agricultural crops, never plants in the wild, but now it seems that wild plants have viroids—in fact, they are the source of viroids that damage cultivated plants. But wild plants are infected with viroids without harm—just as wild ducks are infected with avian flu virus without harm.

Cultivated plants are susceptible to viroid disease because they are a new host, in which the viroid is not so well adapted and must work harder to survive. Cultivated plants are susceptible also because they are often genetically very homogeneous due to inbreeding. Being homogeneous means that if a pest comes along, the whole plant population is susceptible to it—like the lumpers in Ireland. If the population is heterogeneous—a collection of many varieties like the average virus population—some can usually survive whatever happens. With the intensification of agriculture as big business, monoculture is spreading and so are epidemics of viroid diseases. In the short run, monoculture seems preferable because tending one acreage planted with one homogenous crop is easier than tending many small plots that are full of diversity. Also, strains can be selected for their growth rate and resistance to all kinds of pests and conditions. But the very success in growth rate leads to unnaturally high density, which leads to problems among plants as it does among other living things.

Viroids exist only in plants, but a very similar particle infects people: the Delta virus. Absolutely unique in the world of animal viruses, it differs from a viroid only in coding for one small protein. This Delta virus can only reproduce in the company of the hepatitis-B virus and seems to have snagged that one varying protein from a host cell. In short, only by forming a team with a legitimate virus and by robbing a cellular protein could something like a viroid survive in human or animal cells. In their pure form, that is, loose RNA molecules, viroids are apparently unique to plants.

The viroids that plague today's agricultural crops are descendants of the viroids found in a few wild plants. There are, for example, solid indications that the *Mexican Papita viroid* is the precursor of the viroids that now infect cultivated tomatoes and potatoes, with all their disastrous consequences. The plant that the viroid infects in the wild is native to Mexico and has no adverse effects from the infection. The Papita viroid was probably transmitted by insects from this wild plant to tomatoes that were being cultivated in nearby fields. We can only guess how potatoes were infected. They lie under ground but have plants above ground, so perhaps insects reached them with the viroid, or possibly the infection was carried on a hoe or other cultivation tool.

Unfortunately, by now the infection is probably spread throughout the world. This is because seed stocks were collected and exchanged by individual farmers and by farming corporations seeking potato strains in Mexico that would be resistant to potato blight. There is extensive worldwide trade in nursery stock for potatoes, and the seed production for potatoes is heavily centralized, which only furthers the spread of an inherent potato disease.

How do viroids reproduce without coding for even one protein? It has been shown that the RNA of viroids contains specific sequences of nucleic acids that can behave like enzymes in specific circumstances. These "enzymes" eat through the RNA through specific pieces known as catalytic RNA, or ribozymes. With the minimal input of these catalytic RNAs, viroid reproduction can take place in the host cell.

The discovery that these pieces of RNA can act like proteins has importance far beyond the humble viroid. It has led to a new concept of the origin of life and its molecular building blocks. Most scientists agree that an RNA world existed on earth before the DNA world that includes all present forms of life. The old RNA world did not know DNA and, in its earliest times, it possibly did not know any protein either. Viroids consist of RNA and they can reproduce without a DNA intermediary, or messenger RNA. In many cases, they do not need any protein molecules in order to exercise certain enzyme functions. Therefore viroids could be relics from the RNA world. Indeed, they could be base elements for the origins of all life.

Some scientists suggest that viroids may have been the base elements used in forming the reproductive mechanism that brings cells to life—whether they are one-celled organisms, like amoeba, or the cells that comprise multi-celled organisms from fleas to humans. They propose a scenario in which primitive bacteria were the earliest organisms. They were infected by viruses and viroids, and at some point these parasites caused a fusion of all three forms, resulting in a new form: the cell.

Unfortunately, the theory could be proven only if someone could produce a living cell by mixing bacteria, viroids and viruses in a test tube. This is not as fanciful as it may sound, and it has fascinating implications. For example, it might seem obvious that hosts existed before parasites, but by the above theory, it was the other way around. What we now call parasites were once able to live on their own, but at some point they felt threatened without hosts, so they "created" them by setting up cooperative relationships among micro-organisms. As a sort of precedent, we have recently learned that certain bodies floating in all mammalian cells were once independent organisms. These are the mitochondria, which are still independent enough to have their own DNA. They were once bacteria that infected the common ancestor of all mammals. Now they are parasites that work for us, using their energy-making function as a sort of generator for every mammalian cell.

The survival instinct of micro-organisms, specifically viroids and viruses, could thus be the founding force for the origin of life as we know it. In turn, the resulting variety of living plants, animals and people has certainly guaranteed the survival of just as many variants of viroids and viruses. This variety has also assisted bacteria, although, unlike viruses and viroids, they are considered living things that can survive on their own. Only certain primitive bacteria, like a mycoplasma that causes pneumonia, cannot survive without a live cell as a nutrient medium.

Whatever really happened at the dawn of life, it seems that the price of survival for viroids, viruses and some bacteria was total dependence on living cells. Although this price seems high, it apparently had to be paid millions of years ago on the primitive earth. Thus the survival of these forms has become linked tightly—even inextricably, so it seems—to the survival of living organisms, from plants to people.

Some plant viruses were apparently uncomfortable in the world of plants and established themselves in the animal world. This seems to be the story behind circoviruses, which are very small animal viruses with a circular DNA molecule. They are so similar in their DNA to the nanoviruses of plants that they simply must have a common ancestor. The two known circoviruses, a hog virus and a parrot virus, show a remarkably great genetic similarity with certain banana and coconut viruses.

How would a plant virus jump to an animal? It might be injected into an animal by a mosquito-like insect that bit an animal after eating an infected plant. The only problem with this is that nanoviruses of plants are transmitted by insects that eat only plants; they never drink the blood of animals. Some day a primeval insect that ate both foods might be found in a fossil, but it seems more likely that parrots or pigs ate infected plants, then were infected with the plant's sap, which entered their blood through small wounds.

The brand-new animal virus, with DNA as genetic material, subsequently maximized its chances of survival in animals by switching some genes with an established animal virus. This was an RNA virus from the picornavirus group (*pico-rna*, or very small RNA viruses) called calicivirus. The recombination between a nanovirus from the plant world and this RNA virus from the animal world made the new circovirus better suited to spread among animals.

How could such recombination occur? Nanoviruses and circoviruses have a DNA genome while caliciviruses have an RNA genome. It seems likely that a third virus, probably a retrovirus, must have provided the enzyme for making DNA from RNA. Retroviruses get their name from a unique enzyme—reverse transcriptase—that can transcribe RNA to DNA. Harmless retroviruses are present in most vertebrates most of the time, as explained in chapter 6. Sometimes they are endogenous, or built into the host genome, and sometimes they are exogenous or circulating viruses. Either way, they can lend their special enzyme to many uses.

Clearly plant viruses and viroids are important in the world of agriculture and beyond, even if they are not a direct threat to humans. Enteroviruses, on the other hand, threaten us in many ways. A family of the picornavirus group mentioned above, enteroviruses are named for the locale of their favored cells: the gastrointestinal (GI) organs.

As might be expected of viruses that head for the GI tract, most are spread by activities related to food: ingestion or excretion. They enter our system with whatever we put in our mouths. They leave with our feces and get into our sewage systems or directly into ground water, ending up in rivers, lakes and reservoirs. From there, they return to humans in tap water, through crops that have been irrigated, and through shellfish.

Enteroviruses can be inside shellfish or on their shells. Diners often eat these creatures raw after removing their shells by hand. Fish also carry enteroviruses but are almost always cooked and not served with a shell or eaten with the hands. Shellfish pose special risks, which is most likely why they are prohibited by Jewish dietary law and also why travelers are warned against eating shellfish in developing areas. Another danger in such areas is salad, because, like shellfish, it is not cooked, and salad greens are hard to wash thoroughly.

Despite all these exotic possibilities, most enteroviruses enter our bodies with food prepared by kitchen workers who have not kept their hands clean. Clearly these viruses are everywhere, but fortunately their infections are generally transitory and run their course with little or no illness. Although most enteroviruses can cause diarrhea, meningitis, encephalitis and even paralyses, such illnesses are rare. In healthy people, enterovirus tends to reproduce inside the body without the host experiencing any trouble; the virus then disappears unnoticed within a few days or weeks. If an enterovirus upsets a healthy person, it is because in the great diversity of its population, that virus just happens to have extra-high virulence.

Even the most notorious enterovirus—poliovirus—is generally harmless. Most of its hosts do not even notice that their intestines are full of polioviruses. Only 5 to 10 percent have a short bout of diarrhea and fever. Meningitis due to the poliovirus occurs but is very rare, and even rarer is paralysis, which afflicts about one in two hundred people infected with the poliovirus (0.5 percent).

Sometimes a person is unusually susceptible to an enterovirus infection. For example, the Coxsackie enterovirus is harmless unless it infects someone already suffering certain kinds of vitamin or mineral deficiency. Named for the upstate New York town where it was first recognized in an outbreak, it is not usually a problem in affluent countries.

In large parts of China, however, it paves the way for Keshan's disease, a serious heart disorder that afflicts young children and their mothers. Keshan's disease occurs only in those regions of China where there is little selenium in the ground and where the food therefore contains a low percentage of this nutritious mineral. Only children with a low selenium percentage in their blood contract the illness, which can be prevented by giving children selenium.

If we did not know better, we would surely conclude that Keshan's disease is caused by a selenium deficiency. But if that were the case, the illness would always occur, all year round, in any children with selenium levels below a specific threshold value. However, Keshan's disease is clearly a seasonal illness, and this points to an infection. Coxsackie virus and other enteroviruses seem to strike mainly in summer, when people are more often in contact with water through swimming or drinking extra liquids to keep cool.

Coxsackie virus of the B serotype has been isolated in a large number of patients with Keshan's disease. And when the Coxsackie B4 virus was injected into a mouse lacking selenium, it caused enormous damage to the heart muscle. On the other hand, it only caused a very mild illness upon injection into a mouse full of selenium. A Coxsackie variant that caused no illness in a mouse with selenium led to serious heart diseases in a mouse with selenium deficiency. As might be expected, when selenium-deficient mice are fed to repair the deficiency, they are less susceptible to infection or illness related to the Coxsackie virus.

Undernourishment of the host seems to assist the virus in two different ways. First of all, selenium-deficient mice, and possibly people as well, become more sensitive to infection. They are more easily infected with the virus. Secondly, once such a host has been infected with a Coxsackie virus, the virus multiplies to a higher level and is also excreted in higher quantity. So the host is many times more ill and also more contagious.

Malnutrition helps enteroviruses to thrive, but this is not true of all viruses. A 1968 study by Scrimshaw for the World Health Organization suggests that in about one-third of virus infections, the virus is actually hampered by malnutrition in the host. About two-thirds of virus infections benefit from such deficiencies.

In mice, both selenium deficiency and a lack of vitamin E increase damage to the heart caused by a Coxsackie virus infection. This suggests that oxidation stress sends the virus a signal to increase its rate of reproduction. And this leads us to suspect that antioxidants such as copper, zinc, iron, vitamin C and beta-carotene offer some protection against this kind of intestinal virus. Copper and zinc have been tested with good effect in mice. However, with mice in the laboratory, we can make sure they lack only selenium. In people living in the real world, it is hard to say whether the Coxsackie virus reacts to a selenium deficiency, a vitamin E deficiency, or to a protein deficiency called kwashiorkor. When people have too little to eat and face starvation, they are deficient in many things, so finding the main culprit is difficult.

Whether or not viruses thrive in malnourished people, we know that viruses do those people no good. Research by Levander and Beck has shown that reduced quantities of selenium and vitamin E in a host lead to mutations in the Coxsackie virus RNA that increase its speed and quantity of reproduction. It almost seems as if the virus reacts to a dying host by changing its RNA so that it can make the most of the situation before the host dies.

When a host is totally fit, an intestinal virus does not need to make such an effort to survive. Like all viruses, enteroviruses will make every effort to survive, constantly adapting to their host. One of them, the poliovirus, has only one host: humans. It spreads around the world in three different ways, depending on host availability and other factors in its environment.

When things are easy for poliovirus—that is, when access for infection is easy—the virus is everywhere. It is endemic, infects the very young, and causes little or no illness. When poliovirus is subjected to more restrictions but not actually controlled by vaccination, it remains generally invisible but will periodically surface in an epidemic somewhere on earth. When the virus is actively controlled through vaccination, it only shows some very rare convulsions.

The poliovirus might very well become the next virus, after smallpox, to be deprived of its only host and thus eliminated from the face of the earth. Besides vaccination, one reason why things have become difficult for the poliovirus is that living conditions have improved for humans. On the whole, even the poorest are not quite as poor as they used to be. In a way, poliovirus fitness for survival seems to have decreased due to our own increase in fitness for survival.

In areas with high population density, poor living conditions, bad sanitary provisions and tropical temperatures, poliovirus is endemic and causes a mild childhood disease. Before 1850, this was the case in Europe and the United States. Nowadays, it applies in particular to developing countries in Africa and Asia. In such areas, every child over the age of four has already had a poliovirus infection, usually mild and unnoticed. Every mother living under such conditions has poliovirus antibodies, generated when she had her own childhood infection. These and other maternal antibodies are transmitted during pregnancy, so her children will be protected for their first 12 to 18 months. The children will then be old enough to handle a mild poliovirus infection and will generate their own antibodies for lifelong protection.

Ironically, since these early infections often show no symptoms, it was once believed that poliovirus never infects people in poor and tropical regions. On the contrary, it infects more people there than anywhere else. It is simply overlooked.

Poliovirus infections contracted in childhood protect against polio infections at a later age. Primary infections occur less often at later ages but have a much more serious outcome when they occur. The disease was named "infantile paralysis" when its victims were mainly children in areas where the infection was rare but serious. As living conditions improved between 1850 and 1950 in affluent areas, adults were increasingly infected. More and more polio epidemics occurred, particularly in summer. The problem was that, given better conditions, fewer children became infected and thus ever more people had their first exposure to the poliovirus at a later age. Many viruses, including the poliovirus, do more damage when the primary infection occurs in an adult instead of a child. Poliovirus had become hard pressed by the improved host conditions, and its opportunities for spreading had been restricted.

It should therefore not be cause for amazement that poliovirus epidemics struck with more and more frequency in Europe and the United States between 1850 and 1950, inflicting ever greater numbers of victims with serious paralysis. And it was increasingly teenagers and young adults who had these effects. The summer of 1916 was infamous. In the United States, as many as 6,000 people died from polio that year and 27,000 were left paralyzed. In New York City alone, 2,000 died and 9,000 were paralyzed.

The ease with which a virus can spread seems to be inversely proportional to the number of people who are at risk of contracting the infection. The fact is, in a developing country, all people over the age of five have been in contact with the poliovirus and carry anti-polio antibodies. These protect them against the most serious consequences of the poliovirus for the rest of their lives. There are very few adults who can be infected by the virus even if they are repeatedly exposed to it for decades.

In prosperous countries, the reverse is true. The virus needs specific circumstances in order to surface but, once it does surface, the majority of the population, young and old, is susceptible to the virus. In the early 1950s, the number of people worldwide who contracted polio exceeded half a million. In the United States alone, 20,000 people contracted infantile paralysis every year. Many of them were young adults. Before polio vaccination became commonplace in the United States, the last polio epidemics took place among the upper middle class. We are now seeing this story repeated in developing countries, where the improvements in sanitary conditions are leading to epidemics of polio in its worst form.

However, we are on our way to banishing polio by slowly but surely closing off all routes by which the virus spreads. The containment of polio began in 1949, when Enders discovered that the poliovirus could be replicated in a test tube filled with human embryonic cells. This made it possible to produce the virus in quantities sufficient to make vaccine.

In 1954, Albert Sabin published a trail-blazing article in the journal *Science*, in which he demonstrated that cells which were infected with harmless poliovirus variants provided protection against an infection with the true poliovirus. Sabin then had the idea to infect children with weakened polioviruses, thus keeping the stronger poliovirus out. His vaccine does not require injection with a hypodermic needle; it can simply be eaten with a sugar cube. The vaccine can be stored in a refrigerator for a long period of time, because it does not have to be deep-frozen to preserve its potency. The weakened virus behaves like a poliovirus, quickly settling in the GI tract. The vaccine needs to be administered only once, since it infects and reproduces just like a true, aggressive poliovirus.

As Sabin had predicted, GI cells infected with the vaccine strain cannot be "superinfected" with a true polio strain. This phenomenon, known as superinfection interference, is very important in the world of viruses. It means that the weakened poliovirus settles in the intestines of all, or many, people who are at risk of a polio infection. There it stops the spread of the true poliovirus within a few days, even before anti-polio antibodies are made in the body of the vaccinated person. Antibodies are not needed to fight the strong virus because the weak virus will keep it from ever making an invasion.

It is this poliovirus, the Sabin polio vaccine strain, that now dominates in the human population. It survives at the expense of the more virulent poliovirus. But the Sabin strain can cause polio, although with far less frequency and seriousness than the true poliovirus. In 1968, almost 500 children between the ages of 4 and 8 contracted infantile paralysis in Poland, due to the spread of a vaccine strain that had been used 4 months earlier to vaccinate 8 children.

Such an outbreak indicates that the strain used as a vaccine had difficulty spreading. Under such stress, only the most productive strains survive. Of these, it seems that mutations are selected that endanger the nervous system. These viruses penetrate cells of the nervous system, resulting in symptoms of paralysis. Likewise in the true poliovirus, the "neurovirulence" that results in paralysis is apparently a by-product of mutations that help the poliovirus to spread better under stressful conditions.

In addition to this problem, the Sabin strains mutate with some regularity back to their true, aggressive form. These have caused cases of polio in Egypt from 1983 till 1993, in China in the early 1990s, and in the Caribbean region in 2000. But the risk is increasingly smaller than it used to be. The world saw somewhat more than 35,000 polio cases in 1988 but not even 3,000 in 2000.

A handful of countries continue to record polio cases every year. The most important pool is in India, and many cases outside of India, such as in Bulgaria and China, are the result of imports from India. Regions where wars are waged, such as Afghanistan, Angola and Sudan, likewise continue to be a source of polio. The persistence of the poliovirus presents us with the question of what to do when the last polio cases have been controlled with live, weakened poliovirus vaccine.

The current idea is that we will then stop that vaccination and proceed for a number of years to vaccinate with a dead virus vaccine, as was discovered by Jonas Salk. This has proven to be successful against polio in countries like The Netherlands, Sweden, and Finland. The question remains whether it is sufficiently effective and can be administered widely enough to exterminate even the last poliovirus in areas of higher endemicity.

We have managed to eliminate smallpox, but that virus always shows itself by producing lesions on the face or body. Polioviruses are not as lethal as smallpox but may be harder to eradicate, since they so often spread without signs.

My own prediction is that in the foreseeable future, we will never have a generation growing up entirely free from the poliovirus, but there will indeed be far fewer cases of polio than fifty years ago. As a result of a coordinated antiviral vaccination campaign, a new and very favorable balance has come about between poliovirus and human. It is no longer human who is crippled by the poliovirus, but the poliovirus that is crippled by human.

3

RAISING CATTLE AND EATING MEAT

Rinderpest, Measles, and Mad Cow Disease

World population will rise by two billion over the next hundred years and will spike highest in the poorest areas. All these people must eat. For thousands of years, the need to feed more and more people has been met with agriculture and animal husbandry. They are still crucial to our wellbeing, and science is finding ways to increase yield—but these activities both have inherent dangers, which can only grow with their intensification. The last chapter focused on the risks of farming, especially the monoculture favored by agribusiness. Unfortunately, these risks are at least matched by the risks of breeding animals.

Animal protein is an essential part of practically every diet, from north to south and from east to west. Until recently, however, much of the world considered it a rare treat. They got by with protein in small amounts, by choice or necessity, and ate meat in large quantities only on special festive occasions. Nowadays, at least in the affluent western world, most people expect to have meat every day in generous amounts. As the media spread western values to developing areas, the poor want more meat as well—even though some of their high-fiber diets are actually more healthy than those high in animal fats.

Chickens, pigs, and cows provide by far the largest part of the meat demanded by the world. The food chain for human beings can no longer be formed without the breeding and slaughter of these animals. The breeding of livestock is ever more intensive, but this is not healthy for animals. Many animals are kept in crowded conditions that restrict their movement and allow rapid spread of disease among them. This is only partly countered by the widespread practice of mixing animal feed with antibiotics, which are now creating strains of resistant pathogens.

Not only is today's lifestock more in danger of disease, but they are a more dangerous source of disease for other animals, including humans. Two types of animal-related disease have been big news in the past few years. One is foot-and-mouth disease, caused by a member of the PicoRNA virus family, to which polio and coxsackie virus also belong (see chaper 2). The other is the even more frightening "mad cow disease" caused by the mysterious prion.

Rinderpest is a deadly form of plague that can occur among domesticated cows, sheep, and goats. Its name, meaning "cattle plague" in German, is not too familiar because the disease has largely come under control, but it stalked the livestock of Europe from the first few centuries A.D. until the 20th century. The disorder is extremely contagious; its symptoms are gastrointestinal inflammation and bleeding, and it kills most of the animals that contract the infection. In fact, the virus is as deadly for cattle as HIV is for people, and rinderpest kills in a few weeks or months instead of taking several years like AIDS.

Until recently, this disease was rampant in large parts of Asia and Africa, where it had disastrous consequences. It was probably introduced into both Asia and Africa by colonizers from Europe. We know it was introduced to Africa in 1880, by Italian colonists. It soon caused death on a massive scale among kudus, deer, buffaloes, wildebeests and giraffes. This led to mortality among lions and hyenas, which preyed on all these hoofed animals. It also spread among domesticated cattle in Africa, causing serious food shortages.

In affluent countries, such diseases as rinderpest have mainly economic consequences, at an impersonal level. In poor countries, where animals are kept less for commerce than for personal consumption, the consequences are very close to home. In very poor areas such as those south of Saharan Africa, a family with infected livestock or crops cannot find alternative food. Their neighbors cannot spare much, if any, and they cannot buy food at the local supermarket, because there is no market and they have no money. America and Europe are many generations removed from the time when many people were so dependent on what they could raise for themselves. In poor countries, such people are still the vast majority. When starvation threatens, they often cannot survive without food aid from the West.

In 1902, rinderpest was discovered to be caused by a virus. Even before that, in the 18th century, Lancisi in Italy and Vicq D'Azyr in France discovered (independently but at about the same time) that a rinderpest epidemic can be contained by killing and burying all the infected animals while also stopping the transportation of cattle that might spread the disease. These techniques are still used today with diseases like foot-and-mouth disease. Finally, between 1950 and 1960, vaccination against rinderpest was introduced. There is now the hope that the virus can someday be completely eradicated, though it still remains in isolated regions of Pakistan and in war zones such as Somalia and Southern Sudan.

Interestingly, when rinderpest vaccination was introduced to Africa, the wildebeest population increased in just a few years from a couple of hundred thousand to a couple of million. The population of lions and hyenas also returned to large numbers. Thus the survival of predators was shown to be indirectly determined by the resistance of their prey to a virus. In other words, a virus that strikes an animal at a lower level in a certain food chain can affect animals at a higher level in that food chain.

Obviously the survival of prey animals depends on the number and fitness of their predators, but apparently the opposite is also true: the survival of predators depends on the number and fitness of their prey. It depends as much, or more, on the presence of food as on the absence of predators.

Rinderpest no longer occurs in the West, but the related foot-and-mouth disease still breaks out in occasional epidemics. It is rather harmless to cows and causes only a self-limited illness of short duration. However, in 2000, cattle herds in England were decimated due to this disease. In itself it was not fatal, but the epidemic could be controlled only by the systematic slaughter and burial of animals in infested areas. Of course, animal transportation to and from England was prohibited as well, and even tourism was hurt. The disaster has now passed — until the next outbreak. So far Africa has been spared from foot-and-mouth disease, as has the United States, where it is called hoof-and-mouth disease. In fact, it has been limited to England and Western Europe — so far.

The rinderpest virus is related to other animal pathogens but most closely to the virus that causes measles in humans. How can this be? The answer lies in the way virus relationships are traced. Although the rinderpest virus is harmless to people, it seems that there once was a time when a precursor of this virus could indeed infect people, and measles is the descendant that still infects people.

All viruses are assigned to families based on their outward form, their genetic structure, and the way in which they enter the body. Closely related viruses can infect animals that are not related at all. A family genealogy can be drawn from the genetic information available for each member, and this family tree tells which viruses resemble others the most. Such resemblances give clues to their evolutionary history.

When a certain bird virus is found to be much like a certain human virus, we can assume that cross-species infection occurred—maybe recently, maybe eons ago. Looking at the results of infection, we can generally tell whether a bird virus infected a human or the other way around. In the case of the influenza virus, the virus is harmless in most birds, which implies that it is well adapted to them. However, it is harmful in man, in which it is still adapting. Thus it must be a bird virus that jumped to humans, and not a human virus that jumped to birds.

When a virus infects two species and causes significant illness in both, the direction it jumped is harder to determine. This is the case with the rinderpest virus in cows and the measles virus in people. But the evidence suggests the measles virus was one of several that jumped to us from cattle thousands of years ago, when we first began to keep herds for their meat and milk. All viruses need a certain size of population to maintain themselves, and the measles virus generally needs a human population of a quarter to half a million. Probably not by coincidence, this was the size of the earliest civilization in Sumeria, about 3000 B.C., in the presumed cradle of the measles virus.

Although measles probably came to us from cattle, there is now no animal pool for the measles virus. It can maintain itself only in human populations and has evolved with a number of strategies to assure this goal. The virus is extremely contagious; it is contracted upon very brief contact or even through the air. In a typical human family, three quarters of the members who are exposed to the virus will contract the illness if they are not protected by a previous measles infection. Once people have had measles in childhood, they are protected against it for life.

The virus does not go into hiding in people or any other animal. For its continued existence, it is entirely dependent on its being spread rapidly among people who have not yet built up immunity against measles. It is mainly dependent on children, because most adults are immune. The quantity of virus in the body reaches its peak fourteen days after exposure and infection. This peak is marked by a skin rash and is also the time when the infected person is most contagious. However, the person is somewhat contagious before showing any signs of illness, so the virus spreads then too.

Measles infections have much more serious consequences in developing countries than in countries of the western world. Undernourishment plays an important role, and vitamin A seems to be the decisive factor. In cases of vitamin A deficiency, which occur with great frequency in poor areas, there is considerable risk that measles will impair vision or even be fatal.

In such areas, risk is heightened also because the severity of measles varies with the way it spreads. When populations are large and contiguous, measles is mainly an infection of young children. When populations are small and isolated, the virus more often afflicts older people. A community invaded by measles for the first time may lose as many as a quarter of its members.

Abu Bakr from Baghdad is credited as the first person able to distinguish measles from smallpox, doing so in the ninth century A.D. At that time they were often confused. Abu Bakr gave measles the name *hashbah*, meaning eruption in Arabic, and he considered it a form of smallpox. In Italy, where the plague was called *il morbo*, measles came to be called by the diminutive *morbilli*. Morbillivirus is now the name of the whole virus family, which includes the rinderpest and canine distemper viruses.

In 1757, the Scottish physician Francis Home used blood from measles patients to transmit the illness to others. This showed that measles is an infection and not an inflammation, since an inflammation cannot be transmitted. An infection is caused by some outside agent, and may or may not involve disease. An inflammation is caused by the body's reaction—or overreaction—against an invader, real or perceived. It always involves fever or other discomfort, if only fleeting, and sometimes involves severe and prolonged disease.

The infectious agent of measles was not known until 1905, when Hektoen discovered it was not a bacterium, but a virus. He proved this by transmitting the disease with blood that was filtered free of all bacteria.

During those first years of the twentieth century, the trail-blazing work by Koch and Pasteur had taught people that many diseases are caused by bacteria. Viruses were scarcely recognized, but researchers saw they could transmit a disease by way of some "organism" smaller than a bacterium. They saw this by using filters with holes so small that bacteria could not pass through.

In 1911, Goldberger and Anderson succeeded in making monkeys contract measles from filtered fluid from the airways of a measles patient. They were even able to transmit the disease from one monkey to another. When such infectious fluid was placed under a light microscope and revealed no bacteria, viruses could be presumed to be present because something too small to see was causing infection. Today's electron microscope, which makes viruses visible, was decades in the future. Dissection of a virus with DNA recombinant techniques was not yet within the realm of possibility.

In 1954, a measles virus was isolated for the first time by Enders and Peebles, using the blood of a child named David Edmonston. This raised the hope of a vaccine. Earlier attempts to protect against measles were dubbed *morbillization*. As with variolation to protect against smallpox, the procedure involved rubbing the skin of a recipient with tissue from the lesions of a donor with measles; unfortunately, these attempts were never successful with measles.

After 1954, a live but weakened measles virus could be produced for use in vaccination. It would cause harmless infection and thereby prevent later infection by the true measles virus. So David Edmonston's aggressive virus was cultivated in chicken embryos and chicken cells. It was then weakened to what is called the Edmonston B strain. Its use in vaccination caused mild skin rash and fever in many children, but this reaction could be minimized by administering antibodies against measles at the same time. It was approved for sale to the public in 1963.

A second live but weakened measles vaccine was approved in 1965. To make it, the Edmonston B strain was cultivated in chicken cells for an even longer period, reducing its ill effects. Called the Schwarz vaccine, it is still used as the standard measles vaccine worldwide.

Measles could very well become the fourth virus—after smallpox, rinderpest, and polio—to disappear from the earth due to human intervention. When it comes to controlling a virus, measles is the ideal case. There is only one type of virus; the infection betrays its presence by almost always causing fever and a distinctive skin rash; an animal pool no longer exists; and there is an effective vaccine.

However, there are major challenges. Since the true measles virus is extremely contagious and spreads very rapidly, its eradication is possible only if at least 98 percent of the population—that is, almost everyone—

gets vaccinated and acquires immunity. In countries torn by war and unrest, measles will quickly resurface due to failure of the health care system. Also, the vaccine must be kept in the refrigerator and given to children within their first year of life, conditions hard to meet in remote or unstable areas.

Furthermore, even the Schwarz vaccine can be fatal in populations so isolated as to have no tolerance at all. This is true of most viruses, but measles provides a specific and terrible example. In 1964, an American anthropology student by the name of Chagnon traveled to a village in the interior of Venezuela in order to study one of the most isolated tribes in the world, the Yanomami Indians. During the following thirty years, Chagnon led some twenty expeditions to this Amazon tribe. One of Chagnon's colleagues was Neel, a geneticist from the University of Michigan. In January 1968, Neel's team traveled to the Orinoco region to study the immune reaction to a measles infection caused by a weakened vaccine strain.

There are some questionable aspects to this undertaking. Two years earlier, in 1966, Francis Black from Yale had vaccinated a Brazilian Indian tribe, the Tiriyos, with the Schwarz vaccine. He reported that the Indians became much more ill from the vaccine than could be expected on the basis of experiences in the United States. It was particularly dangerous to people with anemia or prolonged malaria infections. However, in 1968, Neel chose the Edmonston B strain for his vaccination studies rather than the much weaker Schwarz strain.

At that time, the Edmonston B vaccine was still given to children in the United States, but only in combination with antibodies to minimize the fever and skin rash. Neel gave his Edmonston B vaccine to the Yanomami tribespeople without also giving antibodies. Within three months after these first vaccinations, the worst epidemic in their history struck the Yanomami. Of the many infected, one of every six died.

Chagnon and Neel reported that, in several cases, vaccination led within only six days to serious measles symptoms. The fever attacks of some of the vaccinated people were so severe that no distinction could be made between true measles and these vaccine reactions. The question remains open as to whether the epidemic was caused by the Edmonston B strain because no viruses were preserved for future analysis. Neel's defenders argue that the Edmonston B strain cannot normally spread person-to-person as measles spread among the Yanomami.

In any case, weakened measles virus can clearly lead to serious symptoms under certain conditions, even if the infections cannot spread. The vaccines now used for the eradication of measles are many times safer for people than the Edmonston B strain, whose risks were acknowledged even in 1962 by Enders, who had discovered it.

With tragic exceptions like the Yamomani epidemic, we have long felt free from the measles virus. But this feeling of security was cruelly interrupted when, in 1994, fourteen race horses and a horse trainer died from a measles-related virus that caused an acute infection in the airways. It happened in Hendra, a suburb of Brisbane, Australia, so the virus is called the Hendra virus. The next year, the virus claimed another human victim, who died from encephalitis.

In March 1999, a measles-related virus surfaced in Malaysia which, like Hendra and measles, spread through the air. Called the Nipah virus, it caused illness and death among pigs, but it also jumped to people, mainly hog slaughterers. Though spread through the air among pigs, apparently a Nipah infection could be contracted by humans only if they had direct contact with live, infected pigs. Eating pork from infected animals did not seem to be dangerous. It was eventually determined that both the Hendra and Nipah viruses originated with bats. Large numbers of bats are infected with these viruses but suffer no ill effects.

All these infections in the rinderpest family kill animals that provide us with food. Or they kill those, like horses, that have been otherwise essential to our existence at some point in history. In general, none of these infections directly threatens us, though rare human cases have occurred. Best of all, the possibilities for eradicating the focus of these infections are excellent.

In stark contrast, our prospects are less encouraging with "mad cow disease," or bovine spongiform encephalitis (BSE). A stricken cow can be recognized from neurobehavioral disorders, such as hypersensitivity to sounds, a swaying gait, aggressiveness, frequent falls, and difficulty standing after falls. Cows contract the disease by eating the brain tissue or bone marrow of infected cows.

By nature, cows are not cannibals, but the BSE epidemic revealed that their feed had long been mixed with protein in the form of ground bones and offal from cows, sheep, and other animals. Sheep may be the source of BSE, since those animals have suffered for centuries with a very similar disease called scrapie.

Once BSE was discovered to be spread by adulterated feed, the industry itself was clearly the cause. Using animal remains in feed is now prohibited, and the epidemic seems to be over. In England, which suffered by far the most, the epidemic reached its peak in 1992, with more than 35,000 cases. Five years later, this number had dropped by 90 percent to only 3,500 cases.

Nevertheless, thousands of animals were slaughtered and buried to contain the epidemic, and that was not all. By 2001, some one hundred people in England had died from a new and surprising variant of Creutzfeldt-Jakob's disease (vCJD). Classic CJD is a dementia that appears to occur spontaneously and cause swift death in elderly people. It is rare but can run in families. In contrast, vCJD can occur in young people and is apparently caused by eating BSE-infected beef.

This was a shocking turn of events, but considering the enormous number of cows infected with BSE and the high rate of beef consumption in England, 100 deaths is a mercifully small number. Cases seem to have declined since 2001 after peaking in 2000. As explained below, vCJD occurs only among individuals with a certain genetic quirk. Studies show, therefore, that the meat of one infected cow is likely to cause, at most, two cases of vCJD, although vast numbers of people may have a bite of that one cow. For example, its ground meat could reach thousands of people if mixed with filler for typical fast-food hamburgers.

So the risk of contracting vCJD by eating infected beef is small and becoming ever smaller due to the rapid decline of the BSE epidemic among cows. Researchers are now predicting that, based on an incubation period as long as twenty years, the people who ate infected beef and may someday develop vCJD should never exceed 10,000. This number is terrifying but would be much bigger if BSE were very contagious to people. Apparently it is not.

The BSE and resulting vCJD caused researchers to take a second look at kuru, a disease that once plagued the Fore people of Papua New Guinea. It was completely baffling until studied by D. Carleton Gajdusek. He received the 1976 Nobel Prize for Medicine and Physiology for proving that kuru, like CJD and scrapie, is caused by an infection.

Gajdusek's anthropological interests led him to the highlands of New Guinea in the 1950s. The country is very mountainous, with isolated and highly diverse tribes, villages and languages dotted all over. Kuru was seen in just 160 tiny villages in the eastern highlands; their total population was about 35,000. All were inhabited by groups of the Fore people. In their language, "kuru" means "trembling," and the disease begins with this symptom. It is an aggressive brain disease that generally kills its victims within a year.

Nobody knew the cause, but the epidemiology provided a clue. The disease occurred among children of both sexes and among women, but never among men. The explanation turned out to be that when certain people died, it was the women's duty to conduct funeral rites that included cannibalism. To show devotion and honor, the brain in particular of a deceased person might be eaten or used as a salve. As a result of campaigns against it, the custom has now died out and, and kuru has virtually disappeared.

At first it had seemed like a hereditary disease. But when chimpanzees were injected with brain tissue of a person who had died from kuru, practically all those chimpanzees would show symptoms of kuru within about two years. Then scrapie also was shown to be transmittable to apes. This was of crucial importance because, in 1959, Hadlow had already called attention to the similarity between scrapie in sheep and kuru in people. Subsequently, classic CJD was transmitted to chimpanzees, proving that a general principle was involved here. Kuru was an infectious disease, as were CJD and scrapie. vCJD would join the group later, after the BSE epidemic in the 1990s.

What type of agent caused these diseases after such a long incubation period?

On the basis of a large number of experiments, Gajdusek concluded that a "strange virus" was involved. He called it a virus for lack of a better term and because the mystery agent was like a virus in several ways. It passed through a filter that allows only viruses to pass through; it reproduced in living tissue; it adapted to the host species; and it seemed to exist in "strains" with different degrees of aggression. These characteristics did not suggest bacteria or other known pathogens.

However, Gajdusek called the kuru viruses "strange" because in some ways they were not like viruses. Mainly, he was struck by their enormous resistance to inactivation. They were not halted by ultraviolet light, for example. This seemed to indicate that the agent contained no nucleic acids, the building blocks of DNA and RNA. However, it turns out that plant viroids—which do contain nucleic acids—are just as resistant to UV light.

We now know that the agent causing BSE, CJD, vCJD, and kuru is not a virus but a prion. Containing no DNA or RNA, a prion is simply a protein of a special type. Every human being and every animal has a prion gene (*PrP*), which codes for these proteins.

Every gene in our chromosomes comes in pairs, and the PrP gene is no exception. Due to a mutation, it comes in two flavors: M (for methionine) and V (for valine). It turns out that anyone homozygous for the M flavor (i.e., both genes of the *PrP* pair are the M type) is extremely susceptible to BSE or kuru. People with two copies of V and those who are heterozygous—having one M and one V flavor—are less susceptible. Worldwide, the vast majority of human beings are heterozygous (or homozygous for V), since the susceptible M/M homozygotes are apparently slowly but surely dying off.

Normally, products of the prion gene do us no harm. In fact, we do not know what they do, or why they are produced. Evidence suggests they were originally used to produce our cell membranes, but now they seem jobless—and some are "sick." These abnormal prions are very harmful indeed.

How are they abnormal? The genetic recipe for any protein includes not only its amino acids, but also its configuration. A normal prion is folded in a certain way, but "sick" prions are misfolded. One result of the misfolding is that such prions cannot be broken down, or recycled, by the enzyme that usually does this job, proteinase K. So the prions accumulate in the brain and cause disease.

The most frightening and baffling thing is that when sick prions bump into a normal prion, they make it sick as well. This can happen when our own abnormal prions contact our normal prions. It can also happen if we ingest a prion from a cow with BSE. Much work remains to be done before we truly understand the mechanism.

We do know that people who are homozygous for the *PrP* gene do not invariably get prion-related disease if they ingest a sick prion, unless perhaps they eat a lot of contaminated meat. They are only more susceptible. But if sick prions are injected into the brain of laboratory animals, they will invariably get the disease. We do not know if this is true also with humans, because nobody would deliberately inject prion-diseased tissue into a person's brain. However, cases of CJD are suspected to have resulted from injecting people with human growth hormone that happened to contain sick prions.

One of the great mysteries of medicine is how genetic mutations interact with external factors to result in disease. Clearly there is an hereditary component in CJD, vCJD, and kuru. In England, 37 percent of the population has the mutant prion genotype, or M/M homozygosity. In Germany, 45 percent has it. Taking the positive view, this implies that more than half of these populations do not have M/M homozygosity. Thus they can probably never contract mad cow disease, even if they should gorge themselves on infected beef.

Perhaps what we see operating here is evolution and selection. Since prion diseases occur in particular or exclusively in people with the mutant prion genotype, the group with a natural resistance against prion disease will tend to grow. Already it is probably much bigger than it was when the mutation first occurred. There may come a time in the far-away future of human evolution when everyone is resistant to prion diseases. Those not resistant will have died off.

Before we knew about mutations in the prion gene, we knew that there are various "strains" of infectious prions. They differ from CJD to vCJD to kuru and also within those three diseases, as to their incubation period and in the severity of symptoms seen in people and in laboratory animals. In a patient with signs of CJD or vCJD, a look at their sick prions can tell us which disease they have: whether it was spontaneous or caused by eating infected beef.

It has been shown that when mice are infected with a slow CJD strain, it protects them against a strain which causes disease rapidly. Among mice who were first infected with a weakened strain and later with an aggressive strain, all contracted only the delayed form of CJD. The mice of the control group, who were infected only with the aggressive CJD strain, all died quickly. This finding is very encouraging because it hints that prion-related disease might be fought by vaccination. Another hopeful recent finding is that antibodies to sick prions protect against prion disease when injected into small laboratory animals.

Strikingly, the brains of the mice infected with only the aggressive CJD strain contained about ten times as much prion protein as those infected with the slow strain. Yet the brain tissue of those with the aggressive strain was 10,000 times as infective, when injected into another animal, than was the tissue of those infected by the weak strain. This was not a complete surprise because researchers had shown earlier that infectivity of brain tissue does not always correspond with its burden of prion protein.

The prion is certainly related somehow to diseases like CJD, and it may be responsible. However, at this writing, no one has fulfilled all four of Koch's postulates to prove this unequivocally. In the 19th century, the

German scientist Robert Koch mapped out four steps to such proof. One, the suspected micro-organism must be present in every case of the disease that is studied. Two, it must be isolated and grown in pure culture. Three, the pure culture must, when inoculated into a susceptible animal, reproduce the disease. Third, the micro-organism must be observed in, and recovered from, the diseased animal.

In the laboratory, nobody has yet been able to induce an infection like CJD by using purified prion protein (or prion protein produced by means of recombinant techniques). So although most scientists consider it the infectious agent, there is still room for debate.

Assuming that it is the infectious agent, it seems to be a unique agent in that each of us carries its makings in our own body. Our own genetic material contains the DNA which codes for the prion gene, source of the potentially sick prion. And every day we are producing prion proteins, albeit the harmless type.

But the most amazing thing about prion protein is that it can take on several stable structures, depending on circumstances. As an infectious agent, it seems to live in the most extreme form of symbiosis with its host. The boundary between parasite, virus and host has become totally blurred. The prion gene does not deviate in any way from the other genes of humans or animals. And it definitely does not show any resemblance to a viral gene. Yet when a spontaneous mutation occurs in the prion gene, the gene product—the prion protein—is misfolded at some point. Due to this change, the proteins coagulate in the brain and cause the person or animal to decline at a rapid pace.

If a person is homozygous for the M/M prion gene and eats a sufficient amount of sick prions, those prions can lead to changes in the form of all normal prions they encounter in that person. The person's own sick proteins accumulate in the brain, causing disease.

At first glance, it all looks like an infection, because the ingestion or injection of a sick protein leads to an increase—as if an infecting virus were reproducing itself. But, in reality, the increase is based on the conversion of normal to abnormal protein.

The two types of prion protein seem to follow different evolutionary laws. In fact, the sick protein almost seems to defy evolution. Clearly the normal protein follows the rules: like other proteins, it survives as long as its animal species survives, changing as the *PrP* gene changes in response to the environment.

But the sick protein is changed by an unknown trigger for an unknown purpose. It survives on the basis of the abnormal PrP gene, which exists only in families that have the inherited form of CJD. And the prion as infectious agent spreads beyond those families only when healthy people get infected with prion-infested hormones, are operated on with prion-contaminated tools or happen to eat the prion-infected tissue of humans or other animals.

We still have much to learn about prions. We still do not know exactly how prions work and why some people contract prion-related infection while others do not. Stanley Prusiner, who won the 1997 Nobel prize for discovering prions, recently told a science reporter: "Anyone who believes we understand what's happening is mistaken." His Nobel prize was the second to be granted for work on prion-related diseases, the first having gone to Gajdusek 20 years earlier.

The worldwide population is clearly at risk from viruses that spread by way of the food chain. This has been the price we paid ever since humans first tamed plants and animals, but the price seems to be going up. If this is because of the exponential increase of the population, perhaps we have not responded to that challenge in the best possible ways. Instead of producing ever more meat with intensive methods, for example, people who already eat too much meat could lower their consumption. After all, we can choose how much we are exposed to these viruses that attack us because we like meat.

The survival of such viruses as rinderpest, foot-and-mouth disease, Hendra and Nipah does not seem to depend on their spreading among humans. The same is true for the agent that spreads BSE from cows to people. People just happen to get in the way. To these infectious agents, we are merely accidental passers-by. We are infected because we meddle with the natural hosts of these viruses and use them on a large scale as food source.

For this reason, these viruses constitute a largely indirect threat to our survival. But we need to consider a new concept of "indirectly infected food." When food is *directly* infected, it harbors a disease agent that immediately makes us sick. We may eat the food, not knowing it is infected, but the food is ultimately lost to us. It is rejected by the body, or its nutritional value is negated by its poisonous qualities. Usually our misery alerts everyone around us to avoid the infected food, so not many people are affected.

Indirectly infected food is also lost to us, but in a different way. It never gets eaten in the first place, because it is not available. Disease reduces our living sources of food, so we simply never get that food. In affluent areas, people can get other food, but this is often impossible in poor areas. Particularly in such areas, "indirectly infected food" has a wider effect than directly infected food. It affects a much larger number of people. Both the sick and the healthy suffer hunger and run the risk of starvation, as do their livestock.

Clearly such damage to the food chain has its most tragic effects in Africa and in Asia. By the same token, the conquest of a disease like rinderpest has the greatest value in such areas. To date, we have been lucky with BSE. However, the lingering threat of food-related disease may be greater than we think.

Humans would not become extinct if there should be no more meat available for consumption tomorrow. But millions of people suffer hunger or even die in developing areas due to lack of animal protein. Such crises are usually due to droughts, but what if rinderpest, foot-and-mouth or BSE became epidemic in the Western world? This may seem unimaginable, but since September 11, 2001, everything is imaginable. Instead of waging biological warfare with smallpox, a terrorist could spread foot-and-mouth disease or mad cow disease among the livestock of its enemies.

Still, nature may remain the greatest threat with regard to plagues that compromise our animals and our food supply. BSE is still spreading at a low rate and has even reached North America, as evidenced by a case in Canada and another in the United States in spring 2003.

4

SLAKING OUR THIRST

The Cholera Bacteria and Its Toxic Viruses

Billions of years ago, before there was life on earth, there was land, air, and a sterile sea. It was in this sea, so most scientists believe, that living cells arose which would eventually develop into the three life forms: bacteria, archaea, and eukarya. Since viruses are not fully alive, they are not included in these groups, but they depend on them for survival.

This fact leads almost automatically to the conclusion that viruses must have originated from primitive living cells, perhaps bacteria. The opposite was argued in 1924 by d'Herelle, a brilliant early virologist from Canada, who discovered and named bacteriophages. Called phages for short, these are viruses that exclusively infect bacteria. As will be detailed in this chapter, d'Herelle's work with viruses and with cholera was extremely insightful, but most scientists think he was wrong about viruses being the precursors of bacteria. If viruses were already roaming the oceans before any form of life existed, how did they propagate? After all, they can only produce offspring inside a cell, with the aid of cellular proteins. All they are able to do without a host is wander around harmlessly, a mere aggregate of molecules.

It is remotely possible that today's viruses are a remnant of viruses that were once more complete forms of life. However, it seems much more likely that bacteriophages originated from bacteria and that their first function was to transmit genetic information from one bacterium to another. They were helpers, and the first life forms needed all the help they could get. In this early stage of life on earth, bacteria had to adapt constantly to extreme temperatures, rarefied air, and enormous air pressure. The best living conditions for bacteria existed in the sea, where even today we see the bacteria at their most diverse and numerous, along with the phages that infect them. Most viruses on earth are essentially aquatic viruses, or were at one time.

The average liter of sea water contains 10 billion virus particles, which is ten times more than its count of bacteria. There are more of both populations in summer than in winter, because conditions are more hospitable. For the same reason, there are more in coastal waters than in open ocean. They are more numerous in icebergs than in the surrounding water, because ice is more concentrated than water. But whether bacteria in seawater are relatively few or many at any one time, the ratio of viruses always keeps pace. This is because most viruses in seawater are supported by bacteria.

The variation in size and form of these phages is enormous, and many are harmless parasites. However, the most important cause of death in marine bacteria is a phage infection. And the more favorable the marine climate is for bacteria, the more favorable it is for their viruses—and the more often the bacteria die from a virus infection. It is like a feedback loop in which viruses keep the bacterial population within certain limits. This population control can in turn effect the plankton and fish.

Unlike bacteria, the viruses that infect them cannot die, not being alive in the first place. But they can drop to very low levels of reproduction, so that they all but disappear. Their biggest enemy is the ultraviolet light in sunshine, though viruses living in warmer regions become relatively more UV-resistant. Wherever they are, they avoid sunlight as much as possible, so they avoid the surface and thrive best farther down, where the water is relatively cooler.

Many phages are not only harmless to bacteria but assist in their survival, as they must have done at the dawn of life. An infected bacterium has by definition more genes (bacterium + phage) than an uninfected bacterium, and extra genes sometimes convey an advantage. For example, a viral gene might code for a poison that kills competing bacteria and provides the host bacteria with more living space. Or a viral gene might enable the bacterium to enlarge its stable of hosts and thus gain more chances to reproduce itself.

Viruses may carry a toxin that is harmless to the host bacteria but harmful to whatever host the bacteria may infect. This is the case with cholera. Since 1884, every microbiology textbook has stated that cholera is caused by a bacterium, *Vibrio cholerae*. This is true, but not everybody exposed to *V. cholerae* gets cholera. They must be exposed in sufficient quantity and—most crucial—the infecting population of bacteria must itself be widely infected by two phages.

Before it was called *V. cholerae*, Koch called this bacterium the *Bacillus comma*, since it revealed a comma-like shape under the microscope. Already in 1884, Koch made a connection between cholera and water. He stated, "People do not just get cholera. Healthy people do not get cholera unless they swallow the comma bacterium. And this bacterium can only grow in the intestines of a human being or in heavily polluted water, such as found in India." This was true, as far as it went, but even early experiments suggested there was more to the story. For example, Max von Pettenkofer asked Koch to send him a pure culture of cholera bacteria. On receiving it, he drank the entire culture down in one gulp and, miracle of miracles, he did not die. What is more, he did not even become ill. He had doubted that the cholera bacteria could always cause cholera, and he was right—as well as lucky.

For a long time it was assumed that such doubters were simply jealous of Koch's reputation but, as with Pasteur, celebrities and geniuses are not always right. Only a few years ago we learned why Koch's story was incomplete. The cholera bacterium causes disease and death when the infection results in a continual watery diarrhea that dangerously dehydrates the host. The diarrhea is the immediate consequence of a poisonous substance excreted by the bacterium, namely cholera toxin. However, the cholera bacterium does not carry this toxin by nature.

The toxin is coded in the genes of a virus called the cholera toxin phage, or CTXphi. This phage, by itself, cannot penetrate the cholera bacterium because the bacterium offers it no receptor. An appropriate receptor is supplied by yet another phage that must infect the bacterium before CTXphi comes along. Called VPI-phi, this other phage is less fussy about receptors and easily finds one by which it can penetrate. Once in place, it provides entry for CTX-phi. Only then can the cholera bacterium produce cholera toxin.

When von Pettenkofer drank the cholera culture, he must have presumed that it contained no toxin. Perhaps he suspected, as we know today, that *V. cholerae* loses its poisonous quality when grown in the laboratory.

The phage interaction that makes cholera deadly was recently discovered, but phages in general were discovered around the time of the First World War. Without actually understanding what a virus is, an Englishman named Twort noticed in 1915 that a virus was killing the bacteria he was studying. In 1917, d'Herelle got to the bottom of the mystery and named the culprits "bacteriophages," or "bacteria eaters" in Greek. Actually, in coining their name, D'Herelle made a mistake in choosing "phage," because he did not really think these viruses ate bacteria. He thought they simply reproduced at the expense of the bacteria. His main point was that these viruses were harmful to their hosts.

We now think that when life began on earth, phages were largely helpful or at least harmless. The harmful ones evolved later. We think that over the eons, coinfection of bacteria by two different virus populations of viruses has often led to recombination, giving some phages a competitive edge.

Clearly lethal phages have the advantage, so the question is, why do they not dominate every single bacterial culture within a very short time? Experiments have shown that the harmless phages always outnumber the harmful phages, because normally the latter do not even appear on the scene until the phages produced within a colony of bacteria have exploded to a very high number. The harmful potential of a specific group of viruses is triggered and increased by virus production that reaches peak levels very rapidly. With such high numbers, there is great virus diversity, and these many types are crowded closely together. The more viruses of different sizes and species are present in a host, the sooner mutations or recombination will bring a lethal group of viruses to the forefront.

Our knowledge of phages grew so fast during and after World War I that the concept of "phage therapy" arose within a decade. The idea that phages could be used to fight bacteria caused great excitement, not only among scientists but also in the general public. In *Arrowsmith,* a widely read novel by Sinclair Lewis (1925), Martin Arrowsmith is a scientist who discovers that "something" is eating bacteria. He called it the X principle. Toward the end of the novel, Arrowsmith is forced to grant d'Herelle the honor of elucidating the X principle and the mechanism on which it is based, the bacteriophage.

D'Herelle, the father of the bacteriophages, put his ideas on paper in

1926. At the time, he was chief of inspection for health services in Egypt, where he encountered plenty of cholera. He had previously focused entirely on bacteriophages which kill bacteria. But he also saw that some bacteriophages infect a bacterium and simply reproduce; others infect a bacterium and hang around as DNA molecules. It is the latter form of phage infection that is involved in cholera.

d'Herelle could not see the whole picture in those days, but he saw a great deal and showed an admirable knowledge of virology for the time. He demonstrated that a phage must be able to attach itself to the surface of a bacterium in order to be able to enter and propagate. He demonstrated that this required a specific "receptor" molecule on the surface of a bacterium. He showed that phages multiply exponentially over a period of hours. And he observed that there was a direct connection between the outpouring of phage particles and destruction of their mother cell. He saw that a phage could reproduce in a bacterium and spread from one to another only by bursting the bacterium. And of course all this applies not only to one-celled hosts like bacteria but to virus-infected cells in multicelled organisms, including humans and all the other eukaryotes.

In the later years of his career, d'Herelle placed all his bets on phage therapy. Using phages to treat bacterial diseases such as tuberculosis, typhoid, and cholera seemed like the only option in that period before antibiotics were discovered. In 1925, d'Herelle began to study the possibility of controlling plague and cholera with phages. In July of that year, three cases of the plague were diagnosed on a ship that had docked in the Egyptian port of Alexandria. d'Herelle treated the agent, *Yersinia pestis*, with a phage that he had isolated in Indochina in 1920. All three patients were cured after one single injection with bacteriophages.

But when d'Herelle tried his plague therapy in India, the Haffkine Institute in Bombay was unable to cultivate the phages in therapeutic quantities. The reason was that d'Herelle had grown phages in a culture medium containing an extract of hog stomach and cow muscles, but in India, this offended both Muslims and Hindus. In 1926, d'Herelle traveled to Bombay, at his own expense, and developed a medium that was enriched with tissue extract from goats, to which papaya juice had been added. This was not objectionable, but the resulting phages turned out to be ineffective, being an Indochina strain that did not work in India. It turns out that phages are extremely specific as to what species of bacteria they will infect.

From that time on, d'Herelle concentrated completely on cholera. He took a leave from Egypt and went to Calcutta in 1927. In that same year, he conducted his first clinical test in the Punjab, using a "local" phage isolated from the *V. cholerae* culture of an Indian patient. In a case-control study, 74 cholera patients received the treatment with phages, while 124 cholera patients received no treatment with phages. In the treated group, 8 percent of the cholera patients died. In the untreated control group, 63 percent died. Studies conducted independently by other researchers showed less dramatic results, but phage therapy did seem to work.

The effectiveness of the therapy seemed to depend mainly on the specific phage that was chosen for treatment and the aggressiveness of the cholera bacteria that had to be controlled. Working at Yale University between 1928 and 1933, d'Herelle studied the variations in aggressiveness of cholera bacteria. Some bacteria caused very serious and fatal diarrhea. Other bacteria, which looked just as bent and curved under the microscope, caused little or no diarrhea. As mentioned earlier, nobody would solve this puzzle for more than half a century.

In 1933, d'Herelle moved to Tbilisi, the capital city of Georgia, then a part of the Soviet Union. It was here that Georgiy Eliava had founded the Tbilisi Bacteriologic Institute to study bacteriophage therapy. Eliava had become a good friend of d'Herelle's when they were working together at the Pasteur Institute in Paris from 1918 until 1921. The Kremlin decreed in 1936 that his institute would be called the National Bacteriophage Institute; it would become the future world center for research on bacteriophages. Although skepticism about such research was increasing in the US, d'Herelle found the Soviet Union willing to lend a ready ear.

While d'Herelle worked in Tbilisi, he formed the Laboratoire du Bactériophage which produced five phage specimens for commercial use. They were sold by a French company called Robert et Carrière, which was later acquired by l'Oréal. The specimens had names like bacte-coli-phage and bacte-staphy-phage. During the nineteen thirties, the Soviet Union developed many centers for phage therapy. Bacteriophage institutes sprang up all over the country, and it became commonplace to treat cases of typhoid and colitis with phages. Two phages were obtained from d'Herelle's company (bacte-pyo-phage and bacte-intesti-phage) and phages to treat dysentery and typhoid were eventually produced in the Soviet Union itself.

Alas, this burst of activity was very short-lived. In 1937, the first Secretary of the Georgian communist party, Beria, was instrumental in having Elavia arrested and executed. A possible reason is that Beria was jealous of the relationship which Elavia had with an actress.

d'Herelle's Russian period ended during that same year. During and after the Second World War, his phage studies were overshadowed by the discovery of antibiotics. The successful treatment of bacterial infections with antibiotics supplanted the bacteriophage therapy until the end of the twentieth century. However, in the 1980s and 1990s, bacteriophage therapy experienced a revival as many bacteria began to show resistance to antibiotics. As this problem continues to grow in the 21st century, phage therapy seems to have a bright future.

Studies conducted in 1982 by Smith and Huggins showed that bacteriophage injections protected mice, chickens, and young cows against a deadly dose of E. coli bacteria. In 1994, Soothhill demonstrated that bacteriophages were very effective against aquatic bacteria that are resistant to antibiotics, such as *Pseudomonas aeruginosa* and *Acinetobacter Baumannii*. These two species are the most important causes of life-threatening infections that occur in patients with severe burn wounds.

Since bacteriophages propagate in their host, only a single dose is needed for phage therapy. Antibiotics do not propagate, of course, so they must be given repeatedly over a period of time to maintain the desired level in the blood. Phages are highly species-specific: they infect certain bacteria while ignoring most others. They can thus be more targeted than antibiotics, and the problem of administering only one, possibly incorrect, phage can be solved by administering mixtures of diverse phages. The bacterium itself will then select the appropriate phages from the mixture—those for which it has receptors—thus causing its own death. Bacteria can develop resistance to phages, as to antibiotics, but this problem too is mitigated by using phage mixtures.

In the case of cholera, however, we must question whether patients really need therapy at all—with phages or antibiotics. Although half of the patients will die if they receive no care of any kind, only one in a hundred will die if they are simply given massive amounts of water. People get cholera mainly by drinking water that contains the toxic cholera bacteria. As the disease runs its course, the patient is dehydrated by watery diarrhea. But if the water lost by diarrhea can be immediately replaced, the patient does not die but begins to recover in a few days. So the cheapest and most effective treatment in cases of cholera is a continuous supply of water by mouth or, in extreme cases, by intravenous administration. The trouble is that finding clean drinking water and beds for this simple therapy can be very difficult in cholera-prone areas.

What about cholera prevention? The bacterium generally lives in drinking water, and the disease will surface wherever the concentration of toxic cholera bacteria exceed a specific threshold—in drinking water, food, or in the feces of cholera patients. While most bacterial infections can be launched by just a few bacteria, cholera is not contracted without swallowing large quantities. Also, cholera does not spread directly from person to person. For example, a nurse caring for a cholera patient would not get cholera from touching diarrheal fluid, as long as she promptly washed her hands. The cholera bacteria must enter the body orally—on the hands or in food or water. It cannot do this if people keep their hands clean, especially when preparing or consuming food and drink.

The discovery of water as a source of cholera goes back to the time, not so long ago, when the drinking water in Europe and the United States was as dangerous as it is today in developing countries. In London, about the middle of the 19th century, the physician John Snow had been struck by the fact that cholera always broke out toward the end of the summer, when people drank more water than usual. They got their water from the Thames River, which was then used for sewage as well as the drinking-water supply. But outbreaks dropped during summer periods of abundant rain, because people collected and drank more rainwater, which was clean. More important, the rain diluted the Thames water and lowered its concentration of cholera bacteria to less infectious levels. Snow was finally convinced that water was the source of cholera when he observed that the disease occurred much more often along the southern banks of the Thames than along the northern banks. Well-to-do Londoners lived north of the Thames and, for them, the water was filtered. The poor, who lived south of the Thames, did without this advantage.

Cholera bacteria multiply best in sluggish or brackish water. In the early 1960s, Rita Colwell discovered cholera bacteria in the intestines of mussels and oysters in the cold coastal waters of the United States. These shellfish bacteria often lacked the viruses which make the cholera bacterium so dangerous to people. The aggressive cholera strains dislike cold salt water, but they will start to multiply with an increase in temperature.

It was found that the cholera bacteria that produce the deadly toxins multiply best in zooplankton, which consists of minuscule shrimp-like creatures. They eat bacteria but also harbor bacteria, such as *V. cholerae.* One of these tiny little shrimps can carry more than 10,000 cholera bacteria. One small glass of sea water consumed during the time when zooplankton propagates is more than enough to kill a person. Cholera bacteria are found in large numbers in zooplankton of the coastal waters of countries where cholera epidemics erupt.

Cholera follows the coastlines of the world's oceans, as if the currents that support zooplankton were spreading the cholera bacteria and its toxic viruses across the globe. Long before anything was known about zooplankton and cholera, the world's oceans were suspected to spread this disease, because men who worked at sea were often the first victims of a new epidemic. At first people thought the main problem was that ships discharged sewage water in coastal areas. Colwell then suggested that zooplankton were the culprit.

Since zooplankton play a central role in cholera outbreaks, their reproductive cycle can often be used to predict outbreaks in specific areas. However, if two outbreaks occur simultaneously, thousands of miles apart, other factors may be at work. A major candidate is El Niño, or the Christ Child, which involves a mixing of cold and warm currents in the Pacific Ocean. The phenomenon was named El Niño by South Americans, because it arises every four to twelve years around Christmastime. It warms the surface of the Pacific Ocean by .5 to 1 degree Celsius, and this can affect the weather and plankton over a wide area. The trade winds force warm surface water from the seas around Peru in the direction of Tahiti. The speed with which the plankton moves can tell us when and where a cholera epidemic will occur. A direct link has been shown to exist between temperature increases at the ocean's surface and seasonally linked outbreaks of cholera, even as far away as Bangladesh.

The frequency of infectious diseases, particularly those spread by water and air and such vectors as insects, can be heavily influenced by changes in air pressure and the surface temperature of the oceans. Fluctuations in air pressure between Darwin, Australia, and Tahiti are designated the Southern Oscillation. In the 1960s, the Southern Oscillation and El Nino were identified as reflections of the same phenomenon and jointly called the El Nino Southern Oscillation (ENSO). Nowadays, scientists are concerned that ENSO events are increasing, with a resulting increase in weather extremes and, in turn, an increase in the febrile and diarrheal diseases spread by mosquitoes, such as malaria. Local healthcare systems are strained—as well as supplies of food and potable water—particularly since the diseases are accompanied by heavy rainfall, droughts, and storms.

The infections most closely linked to ENSO are probably those borne by microorganisms that spend a large part of their life cycle outside the human body. They spend it mostly in seawater (cholera) or in mosquitoes (malaria), where they are at the mercy of the elements. Particularly with diarrheal diseases, the weather not only helps start epidemics but can keep them going. Heavy rains and flooding can easily overwhelm the shaky facilities in developing areas where, even at best, sewage can barely be kept separate from drinking water.

More than three-quarters of the earth's surface consists of water, fresh and salt. The oceans contain between 10,000 and 100,000,000 bacteriophages per milliliter, and these viruses live off the bacteria that accumulate in the many species of plankton. While we worry about cholera outbreaks, approximately 20 percent of the population of bacteria in sea water dies every day as a result of phage infections. And most researchers agree that the balance among life forms in the oceans—specifically in plankton—requires that phages kill those bacteria and plankton cells that threaten to overgrow all other life forms. Thus virus epidemics in bacteria seem to occur in particular when a specific strain of bacteria needs reining in.

Cholera bacteria cannot escape virus attacks. Cholera epidemics do not occur whenever or wherever the surface water temperature rises. Sometimes there are simply too few cholera bacteria in the zooplankton to exceed a critical boundary, so viruses leave the bacteria uninfected and harmless to humans. If the bacteria exceed that boundary, viruses infect them and make them harmful to us and in some cases lethal phages will exterminate the cholera bacteria.

In short, certain aquatic viruses ensure that *V. cholerae* will make us ill. Others remove it from our water. Still others can actually cure the cholera that has become resistant to our antibiotics. But for all these viruses, or phages, the main job is to maintain biodiversity in the oceans. Our part in this vast system of survival and balance is incidental. The process has been going on in the seas since the beginning of life and will continue as long as the seas exist, whether humans exist or not.

Thus viruses can protect and destroy life at the same time. It is just a matter of which life form we are discussing at a given moment. Of the three life forms—bacteria, archaea and eukarya—the form with the most mind-boggling species diversity is bacteria. This diversity, which enables these organisms to adapt and survive anywhere, is shaped and managed by viruses. And sometimes we get in the way, as with cholera.

Humans and other eukarya can survive without aquatic bacteria and aquatic viruses, but not without water. Aquatic bacteria and aquatic viruses cannot survive without life forms like plankton, and they cannot survive on land. So along with all living things—and viruses—we are permanently engaged in an invisible but Herculean struggle for water.

5

WEATHERING STORMS AND DROUGHTS

West Nile Virus and Others

Many viruses depend on certain climates and environments for their spread and survival. For example, the Flaviviruses are carried by mosquitoes, so they thrive where mosquitoes are plentiful: in rainy jungles or in hot, humid and poor urban areas with lots of tepid standing water. The family name comes from Latin *flavus*, yellow, because the most notorious member of the clan is the yellow fever virus (YFV). The Bunyaviruses, in contrast, are carried mainly by rodents and usually thrive in dryer areas. Their name comes from Bunyamvera, a town in Uganda that hosted the first outbreak to be linked with these viruses.

Both families are RNA viruses. Both are basically animal viruses but can also infect humans. Both are prevalent where there is too much water—or too little—and where resources are lacking to control or escape such conditions. But in the last few years, bird migration, human travel and other factors have brought one of each family to the United States.

In the late summer of 1999, New York was struck by an outbreak of the West Nile virus (WNV). This Flavivirus caused illness in 62 people, most of them elderly, and 7 died. Within a short time, the virus had infected about 10 species of mammals in a local zoo and more than 60 native species of birds. Crows and sparrows especially died in large numbers. This is remarkable, because WNV is essentially an avian virus that might be expected to infect birds without harm.

The birds of Africa and the Mediterranean, where it seems to have originated, are widely infected but often not harmed. Apparently it was not well adapted to American birds, so it had to work harder to survive, reproducing to very high numbers. Thus these birds were more infectious to each other and to any people who might contact them.

Within one year WNV had spread to eleven states on the east coast of the United States, and by 2003 it was found countrywide. It caused few human cases but was known to have arrived in an area when the local crows began to fall out of the trees. The virus has now spread south of the United States and, though still mainly transmitted from bird to human, there have been cases of blood transmission and mother-to-child transmission.

The strain of WNV that struck in the United States in 1999 was quickly found to be most similar to a virus that had struck Israel in 1998. There it killed geese but apparently did not infect people. As far as we know, the first human case of WNV was a woman from Uganda, infected back in 1937. Over the intervening decades, the virus became quite prevalent in birds around the Mediterranean Sea and in the southern part of Africa. It has even been seen in Asia. Some of the WNV-infected birds are sick, but many are not.

As might be guessed from its name, it was once very common in the Nile delta. Studies from the period 1950–1960 showed that more than 50 percent of all Egyptians had come in contact with WNV by the time they were adults. It was found that, in 1968, this virus caused about one-sixth of all fever attacks among young children living in Alexandria.

WNV epidemics take place mainly in summer, when heat and humidity make the number of mosquitoes extremely high. The virus survives mainly in birds, with mosquitoes taking it from an infected bird to one that is not infected. A close relative of WNV, the Japanese encephalitis virus (JEV), is also transmitted by mosquitoes. It survives not only in birds but also in pigs. This brings to mind the influenza virus, but unlike flu virus, JEV and WNV do not travel by air. They rarely infect people and, when they do, they cannot spread from one person to another. Each human case must result from a mosquito that bites an infected bird (or pig) and then bites a human. This only happens when the number of people is extremely large in an area where lots of mosquitoes are carrying the virus.

Why did WNV suddenly surface in America? Previously this virus had existed only in Africa, the Mediterranean and Middle East, and

Southwest Asia. The New York outbreak marked the first time that it was seen in the western hemisphere. Why had the virus enlarged its area of distribution or, to ask a more appropriate question: How did it get the opportunity to do this?

Of all the ways it might have landed in America, the least likely is that it was brought by an infected person. Perhaps it arrived legally or illegally with tropical birds from Africa or the Mediterranean area that were imported for American "bird fanciers." Perhaps it came with infected mosquitoes that hitched a ride on a flight from Africa or Israel to the United States. After all, malaria cases have been reported near airports in Europe, due apparently to mosquitoes that fell out of airplanes when their landing gear folded down.

An alternative explanation is that WNV was brought to the United States by migrating birds. Flaviviruses seem to be fanning out across the globe, thanks to infected birds of all kinds. In regions with a moderate climate, its outbreaks occur mainly in late summer or early fall, when the number of migrating birds is highest. If the virus strikes people, they typically live near lakes and marshes, where migrating birds nest and where there are infestations of mosquitoes with a preference for avian blood meals.

Birds seem to be particularly susceptible to WNV infection just before they start flying to their winter home. It has been shown that getting ready for migration is stressful for birds, and stress makes any organism more vulnerable to infection. This is good for WNV, since the birds are about to travel great distances, allowing the virus to enlarge its distribution area very efficiently. In most cases the virus infection does not impair the bird's ability to survive the migration. Of course, few birds migrate from the Old World to the New World. One example is the wigeon, which breeds in Iceland and Siberia but winters on the east coast of the United States. Apparently its frozen breeding areas make the east coast winter seem warm by comparison. And presumably one or two such birds could have been infected before migration by WNV.

The most logical explanation for WNV's arrival in the United States is that tropical storms blew a few infected birds from their West African habitat to America. It is well-known that herons, egrets and gulls can be infected with WNV, and these birds are sometimes spotted on the American east coast. In any case, WNV has arrived in the New World and is now spreading in a westerly and southerly direction in the western hemisphere. This leg of its journey will be much easier than crossing the Atlantic Ocean. Some 155 species of birds pass by New York during their migration toward southern wintering locations. About 30 of these species follow a route that brings them together in lakes and marshy areas with abundant mosquitoes. Crows are clearly susceptible to WNV, and while some crows winter in southern states of the United States, others go to Mexico, Central America, and even South America. However, WNV has killed many of these birds, so crows that carry the virus to South America must be hardier than the average. The crows must be WNV-infected but still able to make the long journey.

Viruses within the same family can differ in many ways. Among the Flaviviruses, WNV and JEV are basically avian viruses that depend on mosquito transmission and enlarge their distribution mainly by bird migration. Mosquito bites keep the number of infected birds high and occasionally infect a human, but humans cannot pass the disease from one to another, so the threat for humans is relatively small. This is not to say that WNV is a passing phenomenon. To date, it has returned to the Americas every summer and also to parts of Europe formerly free of this virus.

By contrast, Flaviviruses like the YFV and dengue fever virus (DFV), are linked to monkeys, not birds. This makes them more adaptable to other primates, like us. They are spread by *Aedes aegypti,* a different kind of mosquito than spreads WNV and JEV. And if *A. aegypti* gives YFV or DFV to a human, the resulting infection can be passed along, person to person. These two viruses are therefore a larger threat to us than either WNV or JEV.

YFV is from Africa and America and has never been seen in Asia. DFV is mainly an Asiatic virus, with very few occurrences in Africa and America. In the jungle, mosquitos carry YFV from monkey to man. In cities, they carry it from human to human, or humans spread it themselves. Also called "jungle fever," yellow fever strikes mainly during times of extreme rainfall, enormous humidity, and high temperatures. In Africa and South America, such times run from January through March. In the Amazon region, young men in particular are struck by yellow fever while engaged in the tropical wood-processing industry. In Central America, yellow fever decimated the men who built the Panama Canal a century ago. Some species of monkeys become ill from the virus, but others do not. In America, spider monkeys and squirrel monkeys die from yellow fever, but capuchin monkeys are unaffected. In Africa, cercopithecus monkeys and colobus monkeys become infected with the virus but do not become ill.

While African and American monkeys are the animal pool for YFV, the pool for the dengue virus is not known for sure. As with yellow fever, dengue spreads among people mainly during the rainy season through bites from the *A. aegypti* mosquito. Where the DFV hides in the dry season is unclear, but perhaps it stays in the mosquito population. Over the past fifty years, its infections have increased explosively due to the increase in human population and the migration to cities, because the *A. aegypti* mosquito is mainly an urban mosquito. The accelerated urbanization has often gone hand-in-hand with poor sanitary conditions and the absence of a system of water mains. This has led to the storage of drinking water in rain barrels, and there is nothing that *A. aegypti* likes better than standing water.

The yellow fever and dengue viruses not only use mosquitoes for transport from one host to another. They can also reproduce in the cells of mosquitoes, as in the cells of susceptible mammals and birds. However, they are not considered to be mosquito viruses that occasionally infect a bird, a monkey, or a human being. The reason is that Flavivirus infections in mosquitoes occur and spread much less reliably than Flavivirus infections in mammals and birds.

If a mosquito bites an infected animal and sucks enough of its blood, the virus infects the mosquito's intestinal cells. The virus leaks out of the intestines and infects the mosquito's salivary glands. When the mosquito has its next blood meal, it transmits the virus, first infecting the area around its bite. To spread the virus efficiently among mosquitoes, however, each infection of a mammal caused by one single mosquito must lead to the infection of at least two more mosquitoes when they suck the blood of that mammal. Since mosquitoes eat very quickly, they often do not ingest enough virus to continue the cycle.

On the other hand, some Flaviruses cannot survive except in a population of mosquitoes. This happens because they infect the sexual organs of the female mosquito and, by this route, the eggs as well. The virus is then transmitted from parent to offspring. Sexual transmission of Flaviviruses in mosquitoes has also been described. In other words, Flaviviruses survive by going into hiding in a pool of birds, in the case of the WNV, in a pool of monkeys in the jungle (YFV), or perhaps in a pool of mosquitoes (DFV). Mosquitoes are very effective as a means of transport because all age groups and types of people get bitten, without exception. The mosquito is unprejudiced, likes all blood equally, and brings the virus right to the blood cells in which it reproduces.

Flaviviruses are apparently establishing themselves in ever more remote corners of the globe. Their survival seems guaranteed as ever more people are struck by these viruses. With WNV and JEV, human behavior has little or no influence. However, with the YFV and DFV, the growing toll is very largely due to our growing impact on their natural virus habitat.

Three factors seem to promote the spreading of these Flaviviruses: the increase in the number of breeding places for mosquitoes, the influx of large population groups into regions where mosquitoes are rampant, and the migration of people from rural and jungle areas to cities that cannot accommodate them. The explosive increase in urban populations between 1800 and 2000 has led not only to more infections but to an increase in DFV diversity. Once people have been infected by these viruses, they are immune, so the viruses must find ways to keep spreading in a population where a growing proportion of people have immunity. The virsuses have been able to circumvent the threat of extinction by producing ever greater numbers of offspring that are resistant to already-existing immunity. These new viruses are probably somewhat more aggressive than older strains because they must fight harder to exist; thus they cause illness with more frequency and with more severe symptoms. But their diversity and agression ensure the survival of YFV and DFV in the human population, despite increasing human immunity to more traditional strains.

There are viruses that thrive when rainfall is enormous, and there are those that multiply best when there is no longer a drop of water to be found anywhere around. Clearly Flaviviruses are in the first group. Bunyaviruses are in the second group, preferring drought. Actually, since they are carried by rodents, what they like best is prolonged drought followed by lots of rain. During a dry spell, many animals that feed on rodents die of thirst. Then, when the rains come, the rodents breed to great numbers because they are free of predators.

An important genus of Bunyaviruses is Hantavirus, which includes four species. They are basically mouse viruses that never harm mice, but three of the four strains can strike people who are exposed to major infestations of mice. People are infected by breathing in air that is contaminated with mouse urine containing the virus. The Hantaan virus, the species for which the genus is named, was discovered in an outbreak of haemorrhagic fever near the Hantaan River in Korea. It is found only in Asia and causes death in 5 to 10 percent of those who contract it.

Another of the four Hanta species, carried by a different kind of mouse, causes the far more serious pulmonary syndrome. This disease has been seen mainly in the United States and causes death in as many as 45 percent of its victims. Its first known outbreak occurred in May 1993, in the Four Corners region, where the four states of Colorado, New Mexico, Arizona, and Utah meet. Populated mainly by Navajo Indians, this desert area is characterized by extreme drought. Its rough climate makes life very difficult for all its inhabitants, both people and animals. Winters are exceedingly cold, but summers are terribly hot and dry; there is hardly a place in which to take shelter from the sun, which shines with great intensity here.

Since 1993, a few hundred Americans have been struck by this virus, and about half have died. So far, the virus is called *sin nombre* ("without a name," in Spanish), mainly because the Navajos did not want it named after them. In recent years, the International Committee on Taxonomy of Viruses has advised that all new viruses be named with "due regard to national or local sensitivities." Previously, viruses were often named after the people or place in which they were first noted, but this custom may now fade.

In China, more than a hundred thousand people are infected with a Hantavirus every year. Although this virus is less aggressive than the one in America, some ten thousand Chinese on average die annually from its haemorrhagic infection. A Hantavirus seen in Europe is even more harmless; it circulates mainly in Scandinavia, killing only one in every thousand infected people. In Finland, for example, one-third of the people contract a Hantavirus infection during their lifetime. It causes little or no disease.

It almost seems as if there is a relationship between the number of people whom Hantavirus infects and the seriousness of the resulting illness. The more people become infected, the fewer problems they seem to experience from the infection, but when only a few hundred infections occur per year, a higher percentage cause death.

There is quite certainly a relationship between the way Hantavirus affects a mouse population and how it subsequently affects a neighboring human population. These viruses do not spread from one human being to another. They strike people only when the number of mice is so enormously high that contact between people and mice can hardly be avoided –when, for example, nearly every blade of grass in a specific area is coated with infected mouse urine. In 1993, it turned out that among all the species of mice living in the Four Corners region, one of every three deer-mice was infected with the *sin nombre* virus. The virus could thus spread easily among the mice. When this is the case, it seems less likely to infect people—but more virulent (and likely to cause death) when it does happen to infect people.

Conversely, if the virus has difficulty spreading in mice and must resort more often to people, it infects a greater number but with less serious effects. We have seen this in China and Scandinavia. In Panama in 1999, we saw a similar relationship: only 5 percent of mice were infected, while 30 percent of people were infected—but mortality among people was not 45 percent, as among the Navajos; it was more like 25 percent. This is still a high rate, but presumably the virus was not as aggressive among people as usual, because it infected more of them and perhaps became more comfortable with them.

The spread of Hantavirus depends not only on the percentage of mice infected in an area but on the overall number of mice, infected and not infected. A population explosion of mice means that even a low percentage of infected rodents will soak the area with infected urine. The size of a mouse population depends on many factors, but mainly the weather. Due to the rainfall in the Four Corners region in 1992, as well as the mild winter, everything was lush and green in the spring of 1993. The *piñon* pine trees produced more pine cones than ever before, and they are a favorite food of deermice. Typically, deermice bear four to five young each and their period of gestation lasts no more than three to four weeks. They usually breed three times per year, but good conditions can bring a fourth breeding season. If predators do not keep pace—are reduced in number by drought, for example—a true infestation of mice can result. This scenario took place in spring 1993.

The Hantavirus is transmitted through the air from one mouse to another. It multiplies mainly in the lungs of infected mice but is also present in their saliva. Once a mouse has been infected, the animal excretes virus for a few months and thus remains infectious to its fellow mice. The quantity of virus in each mouse decreases after a number of months, because the mouse develops anti-Hanta antibodies. A mouse with Hantavirus immunity is safe from future infections. But the Hantavirus never disappears entirely in a mouse population. There are always new mice being born, without immunity.

At any given time, the more mice in a population, the more mice are

Hanta-infected. The more infected mice, the more virus is in the air and in the environment, and the greater the chance that people will become Hanta-infected. People become very ill from such an infection, while the mice experience no effects. Naturally, the more extensive the contacts between mice and people, the greater the chances that people will be infected. But there is little or no chance that a human being will subsequently infect another human.

Interestingly, adult male mice in particular are infected with the Hantavirus. It spreads by way of saliva, and males more than females are exposed because they regularly bite each other while fighting. Also remarkable is that Hantavirus infections are relatively few in small populations. The virus maintains itself best in large populations of mice, closely packed together, because the crowding leads to aggressive behavior, fighting, and virus spread.

The relationship between mouse and Hantavirus—in which the virus does not hinder mouse reproduction in any way, and the mouse guarantees the continued existence of the virus—has apparently existed ever since mice have existed. Every species of mouse has its own Hantavirus. The virus does not spread from parent to offspring but from adult mouse to adult mouse. The virus does not seem to give benefits to the mouse population, but at least it does no harm. Perhaps carrying this virus makes the individual mouse aggressive, thus increasing its chances of survival. There is no evidence for this supposition, but it seems logical because aggressive behavior promotes the spread of the Hantavirus as well as the survival of the mouse.

The weather is important not only to Hantaviruses and Flaviviruses, but also to the Rift Valley virus. A Bunyavirus that survives in cows and sheep, the Rift Valley virus is very closely related to the Hantavirus but resembles a Flavivirus in many respects. For example, the Rift Valley virus is transmitted by mosquitoes. In eastern Africa, where the virus claims most of its animal and human victims, each outbreak immediately follows a period of heavy rainfall. Such rainfalls lead to the flooding of areas where the mosquitoes breed, which leads to an enormous increase in the number of mosquitoes. The most recent outbreak of Rift Valley fever occurred in Kenya, Tanzania, and Somalia at the end of 1997 and the beginning of 1998. All indications are that this was the most extensive epidemic of this virus since its discovery in 1930.

Since the Rift Valley virus takes a high toll among both people and their cattle, an epidemic threatens the lives of both infected and uninfected people. Those infected with the virus succumb in high numbers to encephalitis, internal hemorrhages, and fever. Those who are uninfected suffer from hunger, because the dwindling livestock provide less milk and meat.

Any viruses whose transmission depends on mosquitoes and rodents are affected not only by rain and drought but possibly also by El Niño. As noted in the chapter on cholera, the warming of seawater caused by El Niño has immediate consequences for life in the oceans and diseases caused by aquatic bacteria and viruses. It also influences rain and other weather conditions that in turn influence diseases associated with land viruses like Flavivirus and Bunyavirus. Other factors, such as global warming, may be linked to unusually long-lasting periods of drought or floods, as well as sudden climate changes. Global warming in winter, specifically at the higher latitudes and at greater altitudes, reduces the size of glaciers and thus the stability of the climate. The climate in turn, specifically through extreme changes such as drought followed by extreme rainfall, leads to infestations by mosquitoes and rodents.

It must be stressed, however, that the remarkable spread of viruses during the last half century and the resulting diseases have human causes as well. The circumstances of any host population play a large part in the distribution patterns of its viruses. Around the world, but especially in the poorest areas, we see a mixure of social and ecological problems: exploding cities, impoverishment of the countryside, disappearing rain forests, and loss of habitat for many species of animals. More people are living in humid and crowded urban areas with poor facilities, in densities that assist the spread of viruses.

In the 1920s, the German researcher Penck made an estimate of the maximum population density per type of climate. A warm and humid jungle can sustain the largest number of people per square kilometer. In second place are locales with a warm climate and dry winters, a warm climate with dry summers, a humid moderate climate, and a savannah climate that is dry from time to time. Much lower population densities are possible in a cold climate, whether or not its winters are dry. Only a very minimal population density is possible in a steppe, tundra or desert climate. Several factors obviously determine the density of a human population, but the combination of high density, tropical warmth, and high humidity seems to be the ideal combination for viruses that depend for their survival on vectors such as mosquitoes and rodents. Under such circumstances, humans too become involved in the survival strategy of viruses, often with serious health consequences.

More and more epidemics among humans have a more serious outcome than in the past. Still, they do not threaten to wipe out the human population—but they can alter its composition, as AIDS is doing. In general, virus infections strike only one or two groups. For example, Hantavirus strikes mainly young men, such as lumberjacks or farmers who work in the fields and come in contact with mice; but mortality is low, and it rarely strikes mothers or children. The West Nile virus hits young people and older people, but lethal outcomes are limited mainly to the elderly, who are past their earning and reproductive years.

Another reason virus epidemics rarely threaten human survival is that most viruses depend more on other animals than on us. Yellow fever and dengue viruses need people or some other species of primate, but most others consider humans as a last resort. This applies to the West Nile virus, a bird virus; the Rift Valley virus, a cattle virus; and even the Hantaviruses of rodents.

Though the *sin nombre* virus can be deadly, it infects few people. A virus is deadly when it has a high likelihood of killing its victim; it kills a high proportion of those who are infected. But while such a virus threatens individual survival, it may have little effect on the human species. The infamous Ebola virus is a good example. It is frightening because we still do not know where it comes from; it causes ghastly effects, and often kills. But there is only a very small chance of contracting Ebola. You must be in Africa at the wrong place at the wrong time. And so far, its epidemics are short-lived and kill relatively few people.

Whether a virus is "deadly" is less important to our survival than a combination of three main factors. First is the age groups it strikes, because though everyone is important as an individual, some groups contribute more to the economy and to reproduction. Second is its infectiousness: how easily it moves from one host to another. Third is aggressiveness: how readily it kills its host.

Ebola hits crucial age groups and is obviously very aggressive, but it is not too infectious if precautions are taken. It moves from one person to another by direct contact with various bodily fluids, and such contact can be minimized if certain traditional customs are discouraged. Likewise, HIV also hits crucial groups and is aggressive (if slowly) and not too infectious if precautions are taken—but sexual contact is extremely hard to discourage.

The Flaviviruses and Bunyaviruses are more infectious than either Ebola or HIV infection, because they use a vector—mosquitoes or mice —that is nearly impossible to avoid, especially when weather favors their spread. We are very fortunate that most of them prefer a non-human host.

6

GETTING LUCKY WITH A FAULTY GENE

Escape from Simple Retroviruses

Simple retroviruses belong to the same RNA virus family as the AIDS virus, but unlike HIV, they never attack humans. Simple retroviruses have only three genes while their complex relatives have more (HIV has nine), but both kinds are called "retro" because of a remarkable trick they can do. All have three genes in common: gag, env, and pol. The first two are structural genes, with gag coding for the inner wall of the virus coat and env coding for its envelope, or outer wall. Pol is the tricky one: it codes for the machinery that transcribes the viral RNA to DNA and splices it into the host cell's DNA.

Until 1970, nobody believed such a thing was possible. In fact, nobody had heard of a retrovirus. But in June of that year, the premier journal *Nature* published articles written independently by Howard Temin and David Baltimore. Working separately, they had both discovered the enzyme that would become famous as "reverse transcriptase" (RT). It puts the "retro" in retrovirus, being the crucial player when a retrovirus is integrated into host DNA. Eventually Temin and Baltimore would share a Nobel prize in medicine for this breakthrough.

Temin had been talking about this idea for more than five years, but everyone said he was crazy. At that time, only one-way traffic had been described for genetic information. Everyone thought that DNA transcribed into RNA, which changed to proteins of all kinds. Nobody thought RNA could ever change "backward" into DNA.

Temin had gotten the idea while working on what were then called RNA tumor viruses, 3-gene viruses that cause cancer in fowl. Looking especially at the Rous sarcoma virus (RSV) of chickens, he became convinced that its secret weapon was an enzyme that changed RNA into DNA. Meanwhile, Baltimore was independently working with mouse leukemia viruses and thinking along the same line.

Suddenly they were ready at the same time to describe this enzyme in *Nature.* Such coincidences are not so rare in science, since more than one person is often working on the same "hot" problem. Ideally, all scientific knowledge is shared in the literature, so many people may start working at the same point toward "the next big thing." However, RT was a particularly trail-blazing discovery. It was not only a unique enzyme that did the "impossible," but it settled a key question about the beginnings of life on earth. RNA had long been seen as the most primitive form of genetic material. But if genetic traffic went only one way, how did the earliest life forms (all based on RNA) ever give rise to the life forms based on DNA? The enzyme RT looked like the answer: a new discovery that had been active since the dawn of life.

In honor of this "retro" enzyme, cancer viruses like RSV were soon given the new name of "retroviruses," instead of RNA tumor viruses. Known only in chickens and mice at that time, they are what we now would call simple retroviruses. But since nobody had yet seen a complex retrovirus, the two categories were not yet needed.

The next question was how these chicken and mouse viruses caused cancer, and why they did so only some of the time. Infection with some of their strains caused cancer, while infection with seemingly identical strains did not. The difference turned out to be an additional gene, found only in the malignant viruses. Now called an oncogene, it was thought at first to be a true viral gene. Then two researchers from California, Varmus and Bishop, found that it was a gene picked up by the virus during its infection of a host cell. At home in its rightful host, this gene is regulated and therefore harmless, but once absorbed into the virus genome, it has no regulation and runs wild. When the virus infects a new host, this gene can cause the infected cell to multiply without end, giving rise to tumors.

In a sense, the oncogene makes a host cell immortal, because the cell can't stop regenerating. Given the evolutionary laws, this must somehow benefit the virus—but how? There is a fascinating paradox here. On the one hand, the infected cell is spurred by the oncogene to unlimited multiplication of itself and the virus. On the other hand, these same cells eventually overrun the host and lead to its death. What does the virus do then? How does it benefit by converting a host animal to an enormous breeding vessel of cells filled with identical viruses?

Most likely the virus can spread more efficiently this way and has already gone to a new host when the old host dies. It happens that these chicken and mouse viruses are spread mainly by vertical transmission: from mother to offspring. The chicken retroviruses are transmitted from mother to chicks by way of the egg. Many of the mouse cancer viruses are transmitted from mother to baby through the mother's milk. Vertical transmission is apparently furthered by a very high virus load, and this can be handily achieved by the conversion of virally infected cells into cancer cells. The host suffers, but its illness and death are not the virus's intention; nor is the dying host a problem for the virus, as long as it dies after the virus offspring have been produced.

This strategy is not confined to Temin's chicken cancer viruses or Baltimore's mouse viruses. It is followed also by the mouse mammary tumor virus (MMTV), a simple retrovirus transmitted through mother's milk that causes breast cancer in mice. Since human beings get breast cancer, researchers began to wonder if the culprit might be a similar retrovirus that infects humans. Could a relative of MMTV infect animals larger than rodents? They found it could, and one example was the Jaagsiekte sheep retrovirus (JSRV).

They subsequently found two human illnesses that are (tenuously) linked with an MMTV infection: diabetes and breast cancer. An MMTV-like virus has been found in the white blood cells of people with diabetes, but its role is totally unclear. According to one hypothesis, it activates immune cells that attack and destroy our own pancreas cells, which regulate sugar metabolism. This seems plausible because, in general, MMTV-like viruses have a gene that codes for a "super antigen." This antigen can prompt T-cells to go into overdrive against innocent proteins of the host's own body.

As for breast cancer, it has been suspected for years that some versions of the human disease could stem from a retrovirus infection. MMTV-like viruses are the logical candidate, and evidence was found very recently that such a virus infects only cells in the human breast. It is connected with a certain very aggressive form of breast cancer. But this evidence was found by only one group and was far from conclusive. Another MMTV-like virus has been associated with lung cancer and with multiple sclerosis, which is not a cancer at all. But again, the evidence came from one group and needs much more confirmation.

When we consider that most mice and chickens are full of simple retroviruses, yet only a few get cancer, it seems strange that people—who are so resistant to simple retroviruses—would get abundant tumors caused by these organisms. One demonstration of our resistance is the fact that many human vaccines are produced in chicken cell cultures. Surely some of these cultures get infected with retroviruses that cause cancer in chickens, yet millions of humans, even children, are vaccinated without ever getting a chicken retrovirus. The crucial question is why not? We can certainly be infected by other viruses carried by the animals around us. Moreover, in the laboratory most simple retroviruses of chickens and mice can multiply extremely well in human cells. In nature, is there something that keeps them out of our cells?

Simple retroviruses are typed according to their shape and named with the first few letters of the alphabet. It seems that people are especially resistant to type C, which includes the cancer viruses of mice and chickens. This resistance is hereditary and found only in humans and certain other primates: the apes and monkeys of Africa and Asia. Unlike these Old World simians, the monkeys of the New World—the Americas— lack the resistance, because they evolved differently.

The resistance probably stems from a faulty and inactive Gal gene. This gene and its product are a bit complicated to explain, but they offer a sterling example of how a disease can sometimes pivot on one gene.

The Gal gene was built into the genetic material of all mammals some 125 million years ago, as we know by looking at DNA changes and knowing the rate at which they occur. But as with most genes, there were some faulty ones that could not do their job. Maybe they were especially numerous in primates, who branched off from other mammals some 30 million years ago.

At about that time, the faulty Gal gene acquired unexpected importance. We can only speculate, but we suspect that a series of epidemics— sometimes lethal—were caused by simple retroviruses, whose tell-tale remnants can be found in the DNA of many species, including humans. Over millennia, these epidemics gradually killed off all humans and Old World primates except those with the inactive Gal gene.

So today, most mammals have an active Gal gene while humans do not. And simple retroviruses occur in all animals with active Gal, but never in humans. What exactly does the gene do—or not do in some primates—that gives it this power? It makes the protein Gal, which is an antigen on the outer wall of cells in most mammals. (Gal is a nickname for "alpha-galactosyl sugar groups.")

Like all antigens, Gal is a generator of antibodies—hence its name, "anti-gen." Usually found on a virus or other stranger to our system, an antigen triggers or signals the production of antibodies and immune cells. When a certain intruder with its antigen has been met and conquered, the immune system is primed to raise those same antibodies and immune cells to fight the same intruder if it returns. These antigen-specific fighters arise and act more quickly against a particular intruder than when it invaded the first time.

It is an amazing system that serves all kinds of organisms very well most of the time. Like any system, however, it is not perfect. And it can backfire or have unwanted side effects. For example, we have autoimmune diseases like rheumatoid arthritis or poison ivy, in which the body attacks itself, or a person with kidney failure who gets a transplanted kidney, but his body threatens his life by trying to reject it.

In animals with a working Gal gene, the problem is that they have Gal antigen on the surface of their own cells. Gal is therefore an insider, and their immune system never flags it as an enemy. No antibodies are made against it, nor should they be—as long as Gal is an insider. But Gal can also be an outsider. It is borne on the outer walls of—you guessed it—all simple retroviruses. Therefore, simple retroviruses can handily infect and sicken all animals that have a Gal gene. They can sneak in again and again, each time causing a serious infection, because the animal's immune system makes no anti-Gal antibodies.

In humans, the situation is different. The faulty Gal produces no Gal antigen, so if any shows up, it is treated as an intruder. In fact, when we are infants, our GI tract is normally colonized by a very common enterobacterium that happens to have Gal on its surface. At that early point in our lives, our immune system churns out anti-Gal antibodies. They don't banish the bacterium but render it a harmless part of our normal GI environment. To keep it harmless, they continually circulate. In humans, about one percent of circulating antibodies are anti-Gal antibodies. And if a simple retrovirus tries to invade, bringing Gal on its surface, these antibodies are ready to mobilize.

This is very fortunate for us because the Gal gene and antigen are all around us in other mammals, such as mice, rats, rabbits, pigs, sheep, cows, horses, dogs, and cats. Also the marsupial kangaroo; also fish, reptiles, and birds. Practically all our pets, as well as most animals we eat, have the Gal antigen on their cells and therefore do not make anti-Gal antibodies. They get lots of retrovirus infections and would give them to us, if not for our resistance.

In the laboratory, if simple retroviruses are inserted in human cells, they can thrive without any problems, but in all the thousands of years that we have been taming animals and breeding animals for consumption, we have never naturally contracted any infections from a simple retrovirus. We now know the reason is anti-Gal antibodies, but how do they function? It has been shown in vitro that anti-Gal antibodies block simple retroviruses from penetrating human cells by attaching themselves to the virus envelope. They cover the virus protrusion that fits into the cell receptor so it no longer fits.

Thus we are protected from the direct threat of infection by simple retroviruses. But long ago they could infect us, as evidenced by remnants of their DNA in our genetic material. Found in all mammals, such remnants are called endogenous because they are inborn. They are insiders while exogenous viruses, which circulate, are outsiders. Ironically, the insider could conceivably harm us, even though their exogenous counterparts cannot. To understand how, we must review how they got where they are.

As discussed earlier (and in the introduction), retroviruses can convert their RNA to DNA and become integrated in the DNA of the host. This happens in the nuclear DNA of whatever cell type the virus happens to infect. They become endogenous viruses—or endoviruses, for short. They then live as long as their host lives, though only in a particular cell type. But if the infection takes place at a special moment, the resulting endovirus lives even longer. This occurs when a retrovirus infects a woman at the very beginning of pregnancy. At that stage, the embryo is a cluster of identical cells, the result of a few replications of the original fertilized egg. This is before differentiation, the stage at which these identical cells begin to take different forms so that ultimately they can do different jobs in the body. If a retrovirus gets into the DNA of a cell in this original cluster, differentiation will take it to every corner of the body. It will wind up in every single cell. In all succeeding generations, this retroviral DNA can then be found in all cells of all host offspring.

The endogenous virus is a sort of stowaway and, being an insider, is viewed as harmless by the immune system. Indeed, it usually is harmless.

In some cases it even has a beneficial effect, protecting the host from exogenous viruses that are closely related to it.

For a virus, the advantage in becoming endogenous is that it no longer must fight to spread and survive. In a way, it becomes immortal. Its genetic code is frozen in time or, more accurately, it slows to a glacial pace. It trades the high-speed evolution of exogenous retroviruses for the much slower evolution of its adopted host.

Why do exogenous retroviruses evolve so fast? The evolution of all RNA viruses takes place thousands of times faster than the evolution of the DNA of living things (or of DNA viruses). This is partly because they replicate far more often. In retroviruses, evolution is especially fast because many copying mistakes are made at each replication cycle. The reason is that RNA replication is managed by RT, or reverse transcriptase (the same enzyme that can convert RNA to DNA, splicing it into our genome). For better and for worse, RT is fast and tricky but also very sloppy—a not unusual combination in nature. At each replication, many mutants are generated along with the others. Some are non-viable and soon die, but if they are viable and able to reproduce, these mutants will survive along with the others, starting new lines of their own.

The result is that every population of retroviruses is a huge collection of viruses, all differing only slightly. We call it a *quasi-species*. For viruses, the enormous advantage of existing as quasi-species is that they can adapt quickly and easily to any changes in the environment. When a population of viruses—or anything else—is homogeneous, it can be completely wiped out by some silver bullet. With the heterogeneous population, or quasi-species, at least a few variants can dodge whatever bullet comes along. In fact, there will usually be a few individuals that reproduce even better in the new circumstances. Even if all but one of the collection is destroyed, that one will replicate, and soon there is a new collection growing bigger every day.

The same thing would happen with humans, except we'd need two —one male and one female—to restart the collection. And the process would take much longer, since humans take years to reach the age of reproduction.

So when we say a virus has developed resistance to a drug, we are really saying that the drug killed the majority but left a few unscathed. The dead majority were the "normal" viruses against which the drug was prepared. The remaining few were "abnormal." They were naturally impervious to the drug, due to some genetic quirk. After the drug has wiped out the normal viruses, this minority has the advantage. It breeds up to a new population in which the great majority are impervious. Meanwhile scientists are searching for a drug to which they hope the new population is sensitive. When they find it and kill the majority of the viruses, there will be a few unscathed—and so on, again and again. This is why, with HIV infection, a cocktail of several drugs is commonly used. And even then, resistance can develop, if more slowly.

To return to endoviruses, they are safe from all this, hidden in host DNA. If endoviruses could think, they could not devise a better way to escape bad conditions and wait for better times to re-emerge—and endoviruses do sometimes re-emerge, even after millions of years of hiding. Certain events can bring them to life if enough of their DNA remains intact and has not mutated too much over millions of years in the host.

It has been shown in various animals that their endogenous retroviruses can wake up and become active when the animal is infected by an exogenous retrovirus that is closely related. The inside and outside virus can begin to exchange genes. This activity can give rise to a recombinant virus which starts spreading even faster than either "parent" ever did, and sometimes this new virus makes the animal sick. An endovirus can sometimes wake up even without help from an exogenous virus, as when the host cell is activated by toxic substances or the unrestrained growth of a cancer process.

We can see the many possibilities by looking at the history of endogenous baboon virus (BaEV). Scraps of this virus are found in the DNA of many animal species, telling us it caused ancient epidemics. It is found whole in every individual of four simian species: baboons, mandrills, mangabeys and guenons. It is found in no other simians or in humans, not even as scraps. When we examine the DNA of BaEV in its four simian hosts, we see it is very well conserved: that is, it has changed very little since its integration. This tells us it was integrated recently, in evolutionary terms: no more than 4 million years ago. In fact, we know that some strains were still exogenous 200,000 years ago, spreading from monkey to monkey. Today, however, it is entirely endogenous, with no strains circulating to cause infection.

The DNA also tells us that there were two kinds of BaEV circulating in primordial times. One virus stayed mainly in the forests and infected forest-dwelling mangabeys and mandrills; the other spread mainly on the savannas, where baboons share their habitat with guenons. It seems likely that 4 million years ago, BaEV represented a real threat for these monkeys of the forests and savannas, perhaps even comparable to the threat HIV poses for humans today.

BaEV has been a partner in development of several recombinant viruses and, long before that, BaEV itself appears to have arisen when two kinds of retroviruses merged together. Its gag gene and pol gene seem to have come from one retrovirus, designated as papio cynocephalus endogenous virus (PcEV), whereas the env gene came from a virus designated simian endogenous retrovirus (SERV). They are "designated" because we have only their remains, from which we have extrapolated their original form. The actual viruses no longer exist.

The evidence suggests that SERV spread actively among monkeys until perhaps 7 million years ago, when it became endogenous. PcEV spread among monkeys until about 6 million years ago, then likewise became endogenous—but while it was still active, it infected a baboon that carried SERV. The exogenous and endogenous viruses then mixed their genes, producing a recombinant virus.

For genes to mix in this way, a host cell must be infected by two viruses at the same time, or co-infected. This can happen if two circulating viruses arrive simultaneously at a receptor. But it is most likely to happen if one is already endogenous, or inside. The potential recombinants then produce their RNA, as part of their natural replication, but by accident a molecule of each one is packed into a new particle. Under normal circumstances, two identical RNA molecules of one single virus get packed.

The resulting virus particle with two different kinds of RNA is called a "heterozygote." These heterozygous particles are released from the host cell and infect new cells. During the next replication cycle, two heterozygotes exchange genes. The result of this second exchange is that the new virus particles have two identical virus RNA molecules—as under normal circumstances. But now the gag gene and the pol gene are from PcEV, while the env gene is from SERV.

This is how BaEV arose, and after causing active infection, even epidemics, it became endogenous and harmless. However, such mergers do not create only innocent viruses. Other SERV descendants still circulate and are far from harmless. While PcEV belongs to the type C retroviruses of mice and chickens, SERV is a type D retrovirus. Some of its descendants are type D monkey viruses that still spread in zoos and ape colonies, particularly among guenons. These simian retroviruses, called SRVs, cause a disease like AIDS in rhesus monkeys, though they are harmless to humans. During the 1970s, epidemics of this simian AIDS occurred in several primate research centers.

A new virus could still arise from SERV, which is present in the DNA of the average monkey at about 200 locations. It is remarkable how intact the viruses still are, although they became endogenous so long ago.

The big question is whether such viruses could be called to life in a way that threatens human beings. Unfortunately, as hinted earlier, the answer is yes. The main vehicle is the transplantation of non-human tissues into humans, a practice that is rapidly increasing as we move into the 21st century. Human-to-human transplantation is preferable but often infeasible, mainly because supply cannot match demand. Intra-human transplantation (e.g., moving skin cells from one place to another in the same person) is likewise preferable but not always feasible.

Thus we resort to xenotransplantation (xeno- meaning "outsider" in Greek), using mainly non-human primates and pigs. Xenotransplantation brings cells of these species into direct contact with ours. This is risky because experiments have shown that, for example, Type D retroviruses such as BaEV, can be activated by bringing simian cells in contact with human cells. Experiments have shown that SRVs can multiply and hold their own reasonably well in human cells.

Animal tissue and parts have been used since the 1600s for repairing or replacing their human counterparts, but with very little success until the last few years. In 1682, the skull of a Russian nobleman was repaired with a piece of a dog's skull and, in the 1800s, the skin of frogs was used to heal burn wounds in human skin. In 1920, a Russian physician named Voronoff implanted simian testicles in elderly men, promising new life for their sex drive. In the 1960s, failing kidneys in humans were replaced by kidneys from chimpanzees. During that same period, chimpanzee hearts were also implanted several times in people. In the 1980s and 1990s, baboon hearts and livers were implanted.

The real success was achieved at the end of the 20th century, with the xenotransplantation of organs and cells from pigs that are purpose-bred to provide them. The dark side of this success is abundantly clear, however. Pigs are full of retroviruses, and porcine viruses are definitely able to keep themselves alive in human cells. Four different varieties of porcine endogenous retroviruses (PERVs) have so far been found in pig DNA. In the laboratory two of them have been shown to infect human cells without any problems.

So far, although a large number of people have received implantations of pig tissue, organs, nobody has yet contracted an infection with a pig virus. But the danger remains and in fact increases as we do more xenotransplantation. Not only are there more surgical events to bring pig and human viruses together, but purpose-bred pigs have more viruses than pigs in nature. This is because inbreeding, which nearly always occurs when animals are bred for human use or consumption, greatly increases the number of viruses built into an animal's DNA.

Porcine viruses could enter people in two ways. They could come as exogenous virus particles, hitching a ride with pig cells transplanted into a human being. Or they could come as endogenous viruses, in the form of DNA code, which might be brought to life by merging with a human endovirus.

Recently a pig virus was found that very strongly resembles a human virus. Of the four PERVs mentioned above, one is called PERV-E. It shows a perfect resemblance to HERV-E, a member of the human endovirus family. Such resemblance opens the way to exchanging genes between the pig virus and the human virus. This can, in turn, lead to a virus that can reproduce and spread—from one human being to another, from one pig to another, or from a pig to a human being. Such a virus has thus risen from its ashes like a phoenix and has given itself new life.

All vertebrates and some invertebrates have the DNA of various retroviruses hidden in the DNA of all their cells. Approximately eight percent of human DNA is retroviral in origin. It contains remnants of several retroviruses that are easily distinguishable. One is HERV-E, which is one of a very large family of retroviruses, all look-alikes of type C oncoviruses. Its members are found in the DNA of virtually all vertebrates: cats and gibbons, koalas, pheasants and pigeons, crocodiles and snakes, salamanders and turtles, toads and frogs. The first of these ubiquitous viruses was discovered in a mouse—murine leukemia virus (MuLV)—so the family is called the MuLV group. It includes the simian virus PcEV, the *GagPol* part of BaEV, and the PERVs of pigs.

How does a virus get a foothold in so many animals, from mammals to reptiles and birds to marsupials? At the dawn of life on earth, the common ancestor of the MuLV group must have infected the common ancestor when it was a very early embryo. This ancestor may have been an amphibious creature that gave rise to amphibians, reptiles and mammals. Over the next few eons, as these animals branched out and grew branches of their own, along went the virus—to the last twig. As a result, each animal now has its own member of the MuLV family. Both host and virus have evolved, and for a long time now, it has been very hard for any MuLV retrovirus to switch animals. The hosts have become too different. But at one point, a mammal virus must have jumped to a bird and, more recently, the gibbon virus jumped to a koala (or maybe vice versa). So the jump is not completely impossible.

In addition to these type C retroviruses, human DNA contains many fragments of two other simple retroviruses. One is HERV-K, a type D virus like the Jaagsiekte virus of sheep, the mouse breast cancer virus MMTV, and the simian viruses SERV and SRV. The other is HERV-L, which resembles a type B mouse virus. Type B and D retroviruses are not found in all vertebrates—only in mammals and birds. In both cases, the common ancestor of the retroviruses must have infected mammals after they had branched off from the amphibian ancestor. The mammals then passed it to certain birds.

All three types—B, C, and D retroviruses—have been around for millions and millions of years. Clearly they can survive as long as their host survives—and that is not all. They are not even completely dependent on the survival of their host species. Jumping to another species would not be easy, but the wide variety within a quasi-species of retroviruses makes it likely that at least one strain could make such a change if the preferred host became extinct. So, as long as one animal species with retrovirus DNA in its genome survives, so does the virus.

But simple retroviruses are more than masters of their own survival. They are pivotal to the survival of viruses of totally different families and may even play a role in the survival of mammals. The most famous study on how one virus helps another was conducted in the 1970s by the Russian researcher Zhdanov. He studied the cells of three species that he had infected with a human RNA virus: chicken cells with measles virus; mouse cells with the Sindbis virus (cause of the mosquito-borne Sindbis fever); and human cells with the tick-borne encephalitis virus. He found that in the genetic material of all three types of cells, there was DNA coding for the respective RNA virus. But since the three are not retroviruses and do not have reverse transcriptase, they could not have converted their RNA into DNA for such integration. The job must have been done by the RT of endogenous retroviruses, which are abundantly present in these cells.

In the 1990s, Rolf Zinkernagel went a step further. He discovered RT activity in cells he was using to breed LCMV, an RNA virus. He found that injecting an anti-AIDS substance could prevent this activity. He postulated that RT production of integrated DNA for the RNA virus might enable that virus to prolong its life or its replicative processes in the cell. Alternatively, it might keep the host cell from dying immediately from the virus infection, even when virus load was very high. Either way, the virus would benefit.

If retroviruses can help other viruses, what could they do for mammals? Sometimes their presence enables the host animal to survive, mainly by protecting against other retroviruses. To give just one example, active (not endogenous) retroviruses appear to speed the aging process of certain breeds of mice. These breeds live only about half as long as other breeds of mice. They show all the signs of old age much sooner: memory loss, cataracts, deafness, and baldness. It happens that these fast-aging mice are full of C-type MuLV-like retroviruses, whereas mice who remain young longer lack these viruses.

All indications are that MuLV is the cause of the premature aging and that slower-aging mice have a natural resistance against MuLV. According to a recent discovery, this resistance is coded by a mouse gene that closely resembles the *gag* gene of the human retrovirus HERV-L. How this gene, picked up long ago from an endogenous retrovirus, still functions and actually guards the mice against MuLV remains a complete mystery.

Beyond such survival aids, retroviruses may actually contribute to the very essence and origin of mammals. Unlike more ancient animals, mammals are not born from eggs. They are born alive and are nourished from the maternal mammary glands, which explains their name. This was a new approach that arose about 120 million years ago. On the face of it, bearing live young may not seem more practical than laying and hatching eggs, but it must offer an evolutionary advantage because mammals thrive in every corner of the earth. They are a very successful family of the animal kingdom. However, the evolution toward the bearing of live young did not take place without obstacles.

Half of the genes in any offspring are from the father, after all, and the mother's immune system must be prevented from rejecting the child due to these paternal elements in the child. Similarly, the placenta behaves like a malignant tumor but must still be tolerated by the mother. Moreover, it must serve two contradictory fetal needs. On the one hand, the fetus must be separate enough from the mother to avoid invasion of life-threatening pathogens. On the other, it must be close enough to obtain nutrition and other support for growth. Thus an ingenious solution has evolved: the placental wall is a thin membrane, but its cells are fused together. Nutrients and oxygen can be brought in and waste products can be taken out through the membrane, but the fused cells block pathogens and the maternal immune cells that would reject the fetus.

What do retroviruses have to do with all this? In humans, simple retroviruses generally lead a dormant existence as DNA, but they are remarkably active in a crucial location at a crucial time: the placenta during pregnancy. After the baby is born, they return to their dormant state. Why?

As early as 1988, Erik Larsson of the University of Uppsala, in Sweden, suggested that those retroviruses provide for the fusion of placental cells. Finally, in 2000, the first evidence for this theory was found. Sha Mi of the Genetics Institute in Cambridge, Massachusetts, demonstrated that an envelope protein of the human endovirus HERV-W (a type-D virus like MMTV) is essential to formation of the placenta's fused cell layer. This protein is called syncytin. It is produced by a retroviral gene that humans long ago added to their DNA. The retrovirus used syncytin to fuse to a host cell before penetrating and causing infection. Now humans use this "sticky" protein to fuse the cells in the placental membrane. It also brings a bonus in that, being part of the retroviral transmembrane, it has immune-suppressive qualities that could help ensure, inside the placenta, that the mother's immune system does not attack her child.

Mi's study seems particularly convincing because whereas syncytin is coded by the retrovirus env gene in the host DNA of all of the body's cells, it is expressed or activated only inside the fused placental layer (and to a very minor extent in the testicles).

This research indicates that a retrovirus has played a role in the formation of the human placenta—the key to mammalian life. Without this layer, the placenta does not function and, without a placenta, offspring cannot be born alive, which is to say, mammals would not be mammals. The question remains whether this has been the case in all mammals.

So simple retroviruses once infected primates as well as other mammals and birds. Their survival strategy was to link their fate to that of the host as endogenous viruses. They even created their own "eternal" host: placental mammals. Meanwhile, humans and some primates became resistant to these viruses in their exogenous form. We therefore avoid not only the infections they cause in other mammals but the cancers that can result from those infections.

However, we still carry remnants of endogenous simple viruses, and our increasing practice of xenotransplantation could bring these viruses together with other viruses to form a new threat.

7

TAKING CHANCES WITH SEX

The Herpes and Papova Viruses

Humans and many other species of mammals live intimately with two virus families that have a lot in common: the herpes and papova viruses. Both have DNA as their genetic material, as do their hosts. Both are very specific to their host species, so they do not jump around from one species to another. Both are transmitted by the most intimate bodily fluids, including those involved with sexual relations. Finally, both can hide in their host for a long time without causing trouble, except for occasional episodes—but some of both families can ultimately cause cancer.

At this point in the HIV era, everyone has heard of the herpes viruses. There are more than one hundred herpes viruses in the world, each one infecting a certain animal species. Humans must contend with at least eight.

Of these, I will focus on herpes simplex virus (HSV), which causes painful but fleeting sores; the Epstein-Barr virus (EBV), cause of mononucleosis; and the human herpes virus 8 (HHV8), cause of Kaposi's sarcoma. These three viruses do not have an animal pool and are thus completely dependent on the human host for their survival.

Herpes viruses have a much larger genome than most viruses. Composed of single-stranded DNA, it is at least ten times longer than, for example, the AIDS virus. HIV has nine genes, and the herpes viruses have dozens more than that. Many herpes genes have been hijacked from a host cell, which retains its own copy. These shared genes make it easy for the virus to communicate with the host cell and to survive there in a latent form.

Herpes viruses are life-long parasites, but they do not integrate into the host DNA. They hide somewhere in each host as long as he or she lives, which is apparently a prerequisite for their survival. They age at the same rate as their host. In some herpes species, this parallel aging and close interaction with the host cells sometimes lead to cancer.

Another characteristic common to all herpes viruses is that they are transmitted through intimate contact, usually through sexual secretions or saliva. Once inside the new host, they head for their favorite cells. In humans, HSV and varicella zoster virus home in on certain neural cells, but EBV and HHV8 prefer blood cells. All four tend to lie dormant for long periods. They cause discomfort only during acute infection—when they first infect their preferred cell—or during flare-ups. They reproduce only at those active times, typically causing small painful sores. These lesions fill up with large quantities of virus that can easily be transmitted to another person.

The most common herpes virus is HSV, which comes in two variants that greatly resemble each other: HSV-1 and HSV-2. They are virtually twins except that HSV-1 is transmitted by saliva, while HSV-2 is transmitted by sexual secretions. During acute inflection and flare-ups, they cause cold sores or genital lesions, respectively.

HSV-1 spreads mainly among children, with virus in one child's saliva infecting another child. This happens easily in the early years when children drool on everything they handle and routinely put things in their mouths. They might share a spoon when eating, or lick the same ice cream cone. They might scuffle and even bite each other. HSV-1 is so contagious that practically everyone all over the world has been in contact with it. We have all become infected by the time we reach adulthood. Fortunately, though infection is lifelong, its effects are minor: the occasional "cold sore." This is no doubt why people do not worry too much about preventing its transmission. In any case, that would be impossible without keeping all children apart from others.

HSV-2 behaves very differently. The first infections with HSV-2 take place as children reach the age of sexual activity. It is transmitted during sexual intercourse, and the risk of being infected rises with the number of sexual partners. In the various countries of Europe, percentages of adults infected with HSV-2 ranges from 10 to 20 percent. In the United States, the range is greater, reaching 50 percent in some areas. Contagion is particularly high among less affluent people.

Research into the DNA of HSV-1 and -2 indicates that they branched off from the same common herpes ancestor some 8 to 10 million years ago—about the time man branched off from the other primates. Probably the common herpes ancestor used both transmission routes—sexual fluids and saliva—as do some simian herpes viruses today.

The best example is a virus of macaques, the herpes-B virus. In a typical macaque colony, herpes-B infects about half of the individuals when they are very young: long before they become sexually active. The other half are infected during their first year of sexual activity. Using two transmission routes, herpes-B is first a virus of very young monkeys, transmitted through saliva as they eat or tussle together. Then, when mating activity begins, herpes-B acts like an adult virus.

In very exceptional cases, the herpes-B virus can infect humans. Discovered in about 1930, it is the only one of some 35 simian herpes viruses that can infect us. Though harmless in monkeys herpes-B causes death in 75 percent of human cases and was in fact named for one such case. In 1932, a young doctor with the initials WB was bitten by a macaque and soon died from lung failure, after a serious inflammation of the spinal cord. In 1933, the virus was isolated from WB's tissue samples and named with his first initial. About a year later, working independently, Albert Sabin also isolated the virus and named it with the second initial. As often happens in science, the new discovery started out with more than one name. For some reason "herpes-B" stuck, even though Sabin was not yet famous for his oral polio vaccine.

In people, herpes-B first causes an influenza-like illness, then infects the brain, causing the immune system to react with usually lethal effects. (Often what we experience as illness is actually our immune system attacking an invader.) In macaques, the virus goes into hiding in the brain after a very mild acute infection. It maintains a dormant state and becomes active only in stressful circumstances. Even then, it is not lethal. Macaques carry the virus their entire lives in offshoots of the brain, such as the spinal cord, where the virus is somewhat screened from attacks by the host immune system.

The reason that HSV-1 and HSV-2 branched off from an ancestor like herpes-B probably lies in some drastic change that occurred among primates. It may well relate to a fundamental difference between human and nonhuman primates: the way they mate. In humans, mating is powerfully influenced by visual cues between potential partners. During the mating act, humans are usually face-to-face. However, in nonhuman primates (and most other animals), visual cues are less important. Mates are chosen mainly because they are handy during the fertile periods that regularly occur due to hormonal cycles. Females become receptive and males respond. During the mating act, simian partners are not face-to-face and do not exchange saliva by kissing.

Possibly, when humans began to adopt their way of mating, the ancestor virus changed to survive. From being a virus transmitted by both saliva and sexual secretions, it changed into two viruses, each a specialist in one of these routes. Put another way, variants of the ancestor that happened to be particularly good at transmission by saliva survived as HSV-1; variants particularly good at sexual transmission survived as HSV-2. All the others died off.

The herpes simplex viruses specialize in a way that is ingenious but quite straightforward. Both viruses infect neural cells and spend most of their time in dormancy. But when they are activated by stress, or some other reason, HSV-1 typically surfaces around the mouth. It causes "cold sores," which are so-named because the causal stress is often a "cold" or some other burden on the immune system. On the other hand, such stress causes HSV-2 to surface around the genital organs. The result is genital herpes. Like cold sores, the genital sores are painful and relatively untreatable, but they quickly disappear.

It should be noted that while genital herpes is caused mainly by HSV-2, it can likewise be caused by HSV-1. For this to happen, the infection must occur sexually instead of by saliva. This rarely happens because by the time most people are sexually active and might be sexually infected by HSV-1, they have long carried dormant HSV-1 from a childhood infection. Early HSV-1 infection protects against HSV-1 infection later in life. Unfortunately, neither early nor late HSV-1 infection protects at all against HSV-2 infection.

The Epstein-Barr virus (EBV) and Kaposi's sarcoma virus (KSHV) are a completely different story. In each case, there are times when the host cell cannot satisfy the requirements of the virus except by giving itself eternal life at the expense of the host. It begins to replicate out of control, resulting in cancer that can sometimes kill the host.

When EBV was discovered in 1964 by Epstein and Barr, it was not linked with cancer. In fact, it was not linked with anything until 1968, when it was shown to cause Pfeiffer's disease. In 1889, Emil Pfeiffer from Wiesbaden had described a condition consisting of enlargements of certain glands plus symptoms like fever and fatigue. We now call it mononucleosis because, in a person with this infection, the circulating blood shows a dramatic increase of T-cells called mononuclear leukocytes, especially of an abnormal kind. However, as we will see, mononucleosis is the elite version of EBV infection.

EBV is everywhere around us. It is spread by saliva like HSV-1, but requires more intense exposure to cause infection. The infection occurs earlier or later in a person's life, depending mainly on sanitary conditions. If such conditions are poor, with people crowding together and eating with their fingers, EBV infection is early and very mild. By the age of 5, almost every child in Africa, Asia and South America has been infected with the virus, but most of them have not noticed.

In the more affluent Western world, only 50 percent of 5-year-old children have been in contact with EBV. As they become teenagers, they may get Pfeiffer's disease, or mononucleosis. It is an illness of the affluent Western world because it is spread not by poor hygiene or living conditions, but by deep kissing. EBV infection in teenagers or older people is more severe than it is in the children of developing countries. Its symptoms are relatively mild, but they include extreme fatigue, and a person with "mono" may need several weeks to feel well again.

Whatever the form of EBV infection, EBV stays with the infected person for life. After the acute phase, the virus never reactivates under normal circumstances. It gives no outward sign of its presence, but testing finds it in approximately one out of every hundred thousand B-cells in the host. These immune cells, which make antibodies against intruders, are found in the circulating blood and also in the spleen, in the lymph nodes and in the tonsils. We carry EBV around with us in these organs, and it cannot escape. Nor can it be recognized by the immune system, as long as it stays in these organs. Small bits of virus regularly leak from the infected cells, but the normal immune system is easily a match for these small quantities. The virus never gets very far when it leaves its cell—only up to the organ's edge, at most.

EBV infection is very widespread but often quite harmless, at least under normal circumstances. But if resistance decreases, for example, due to infection by the AIDS virus or due to medications after an organ transplant, the virus may reactivate. EBV only infects very few cells, so few that

EBV survival is jeopardized. It may then cause an unrestrained growth of the EBV-infected B-cells. The virus, for its own survival, needs to increase their number, so it prompts their uncontrolled replication. In effect, the B-cells become immortal, and this means only one thing at the cellular level: cancer.

EBV infection can give rise to two types of cancer: Burkitt's lymphoma, which occurs only in certain African children, and nasopharyngeal carcinoma, which is limited to certain mostly-Asian adults. Both types of cancer arise when environmental factors disturb the "normal circumstances" that usually keep EBV quiet. We do not know their exact incidence or mortality rate, because they occur in areas where record-keeping is poor. But they can be lethal.

Denis Burkitt was a British surgeon working in Uganda during the mid 20th century. When he discovered the lymphoma that bears his name, he sensed that climate played a role and suspected a virus or bacterium as the cause. Epstein and Barr later found EBV particles in the tumor cells, but the climate connection is still mysterious. Burkitt's lymphoma occurs only south of the Sahara Desert, in tropical lowlands with a very high level of humidity. It occurs in coastal areas but not savannahs. What some call the "lymphoma belt" extends from countries such as Ghana in West Africa to Tanzania, Uganda and Kenya in East Africa, by way of Congo and the Central African Republic.

No insect carrier has been found, but malaria is suspected to play a role, perhaps by weakening the immune system. The direct relationship between malaria and Burkitt's lymphoma is totally unclear, but the two diseases occur in the same region. Also, it seems that malaria prompts B-cells to multiply, in particular, the B-cells infected with EBV. In so doing, malaria seems to activate the transformation of these cells into permanently dividing B-cells. But we do not know how or why this happens.

Nor do we know why Burkitt's lymphoma occurs in only a fraction of all the African children infected with EBV. In fact, Burkitt's lymphoma is now on the wane in Africa. Perhaps this is a result of efforts to prevent or postpone EBV infection. The infection is postponed in general by an improvement in living conditions and, in particular, by discouraging adults from a custom of chewing food for children. This traditional custom was—and remains—widespread in Africa. Its curtailment will cut down on EBV infection, as will more washing of eating utensils, more washing of hands before and after eating, and by more use of soap in both cases.

Nasopharyngeal carcinoma is also closely related to EBV infection but does not occur conspicuously often in Africa. It affects the nose and throat membranes and strikes mainly Chinese people, but Eskimo inhabitants of Greenland and people in Tunisia are also affected. Though very rare in Japan and North China, it is endemic in the south of China. All three of the susceptible population groups contract EBV infection at about the same young age, but in this they differ little from the Africans who get Burkitt's lymphoma. What makes them vulnerable to nasopharyngeal cancer?

Researchers suspected that diet played a role. The big dietary difference between people living in the south of China and people living in the north is that the southerners eat lots of salted fish, fermented salty soybeans, and dried squid. Salted fish seems to be the most likely culprit, being the only one of these foods fed to babies as well as to older people. Researchers found that the Chinese who later contracted nasopharyngeal carcinoma had rarely eaten fruit or vegetables during their childhood. Other researchers found a relationship between the occurrence of this form of cancer and poverty, Buddhism, and the presence of small altars in the home.

These findings and dietary findings may seem unrelated, but taken together, they indicate that a contributing factor to the contraction of nasopharyngeal carcinoma is the observance of a traditional life style. Such a lifestyle goes hand in hand with poverty and a diet limited to very few foods. Something like salted fish might be harmless until it becomes a major part of the diet. In the other two high-risk areas, researchers found different food suspects—such as dried berries and fish in Greenland and dried meat in Tunisia—but these are all preserved foods, like salted fish. Salting, pickling and drying of food to save it for lean times is traditional among populations in specific regions where conditions of relative scarcity prevail.

Ultimately researchers found that volatile nitrosamines in preserved foods are the substances that turn EBV-infected cells into cancer cells in some people. When it is dormant, EBV hides almost exclusively in skin cells of the mouth and throat cavity, and as the infected cells meet nitrosamines over a long period of time, nasopharyngeal cancer can appear.

As with Burkitt's lymphoma, EBV infection is not the cause of nasopharyngeal cancer, but it must be present for the cancer to occur. Most people in China, including those in the south, are infected with EBV—but very few get the cancer. It seems that the earlier the EBV infection, the greater the risk of both cancers, and early infection is mainly a consequence of poor living conditions. Yet nasopharyngeal cancer appears in adulthood because it stems from long exposure to nitrosamines. In contrast, Burkitt's lymphoma occurs in young children because they are the most susceptible to malaria.

In both cases, the cancer is an unwelcome side-effect of the rapid and effective spread of this herpes virus in a specific population. Several conditions must coincide. The cancers can develop—whether early or late in life—only if the host is EBV-infected very young and faces certain effects of poverty and traditional customs: malaria or a diet with a high proportion of preserved foods.

Kaposi's sarcoma differs in important ways from the cancers that can arise with EBV infection. First of all, it is linked with a different herpes virus: HHV-8. Second, infection with this virus causes cancer directly and does not merely provide a setting in which cancer may possibly occur. Third, the infection is not necessarily an effect of poverty.

The infection and the cancer are rarely serious unless the person is also infected with HIV. In fact, HIV has put KS on the map. Before HIV emerged, few people had heard of KS, and the causal virus had not been discovered, or even looked for. That virus, discovered only a few years ago, is called HHV-8 because it was the eighth human herpes virus to be found. Often it is called the Kaposi's sarcoma herpes virus (KSHV).

Except for the swelling of tumors, most cancers are not visible to the naked eye, but KS is a vivid exception. One of the strangest cancers in existence, its "tumor" involves the unrestrained growth of the cells that make up our blood vessels. Vessels are formed over the entire body, in locations where they do not belong, and can be easily seen as red-pink spots on or in the skin. Instead of metastasizing from a single locus, like most cancers, KS seems to start with metastasis—but it does not go backward to a single locus. It just keeps spreading.

Despite this visibility, KS has historically been a minor health problem in terms of the number affected and the seriousness of their disease. Classic KS occurs in a number of more or less aggressive forms among older men in various risk groups and in various parts of the world. Mostly they live around the Mediterranean Sea or originally came from that area. Some 10 percent of the Mediterranean population—male and female— is infected by the KSHV in young adulthood, but we do not know the exact means of transmission. We suspect it has something to do with traditional domestic customs or hygiene.

Whatever the route of viral transmission, classic KS does not appear until many decades after KSHV infection. The cancer is seen mainly in men over 50 in Greece, Southern Italy and countries such as Tunisia, Morocco and Algeria in North Africa. Another form of KS occurs frequently in Africa south of the Sahara. It is more aggressive and is found more in boys than in men.

The classic KS seen in older men living in the Mediterranean area is seen also in Jewish men living all over the world. Whatever makes them susceptible must predate the Diaspora, when Jews left their Mediterranean homeland two millennia ago. Driven out by the Romans, they went in two main directions and thus form two main groups today. Those we now call Ashkenazi Jews spread up through Asia Minor and into Eastern and Northern Europe. Those now called Sephardic Jews spread across North Africa and into Spain. A number of genetic illnesses occur in just one of these groups, indicating mutations that developed after the Diaspora. KS occurs in both groups, though not in great numbers. When Jews now living in Israel were tested for susceptibility to KS, birthplace was found to be important. The Sephardic immigrants from North Africa appeared most likely to contract infection with the virus.

Though KS is a very old disease, most of the research is recent because it was spurred by the astonishing spread of the virus in the context of HIV infection. In the early 1980s, AIDS was recognized mainly due to the skin abnormalities seen in homosexual men. The abnormalities were later found to be symptoms of KS, so KS was linked with HIV infection. However, not everyone with AIDS has KS. It virtually never occurs in heterosexual men with hemophilia, who were HIV-infected by tainted blood products before they were rigorously screened for HIV. Nor does KS occur in heterosexual men who inject drugs and are infected with HIV (or the hepatitis B and C viruses) through sharing of needles and syringes.

The fact is, there is little or no transmission of KSHV by blood. This is because KSHV is not present in the blood of an infected person in significant quantities, except during the acute infection and at a very late stage of the infection. At both those times, skin lesions are obvious and tend to discourage contact.

The form of KS that can accompany HIV infection is seen almost exclusively in homosexual men. How KSHV might be transmitted in this population was studied in homosexual AIDS patients in Amsterdam. This work revealed that whereas HIV was spread mainly by anal-genital contact, KSHV was spread through oral-genital contact. It seems that

KSHV, like EBV, lies dormant in cells of the throat cavity. It is abundantly present in saliva during acute infection, which often includes symptoms like coughing and a scratchy throat, in addition to muscle pain and fever. The study also found that KS occurred especially in homosexual men who first contracted an HIV infection and then a KSHV infection. Our own research suggests that HIV increases the number of receptor molecules, that allow HHV8 to enter certain blood cells (monocytes and macrophages). More HHV8-infected cells leads to more spread of HHV8 and that in turn leads to Kaposi's sarcoma.

Interestingly, the strains of KSHV found among homosexual men in Amsterdam were traced to the Mediterranean area, specifically Southern Europe. I suppose the virus entered the Dutch homosexual community no later than 1956. Just a few years before that, Dutch companies began very actively to recruit guest workers from Italy and Spain. The first of these immigrants, who came to The Netherlands starting in 1949, were Italians. Many of them came from Sardinia, an island where many people are infected with KSHV.

These guest workers were largely heterosexual, but somehow KSHV entered the homosexual community. It turned out to be transmitted most efficiently by homosexual contact and therefore became almost endemic among homosexual men. A virus can make a change of transmission route with just one incident of novel exposure, if results are good. And through our study of HIV we know that an incident is not impossible among largely heterosexual men. A guest worker might have been bisexual, for example, or might have engaged in homosexual activity for badly needed money. In economically deprived populations, sex of all kinds is often used as a way to earn subsistence income.

However KSHV is transmitted and whatever the form of KS, it seems that the virus infects both sexes but predominantly targets men. Certainly KS favors men. Perhaps a mutation on the Y chromosome makes men more susceptible, or maybe male testosterone heightens their susceptibility. Alternatively, perhaps estrogen is protective of women, or perhaps women one way or the other have fewer HHV8 receptor molecules on their cells.

What could be the advantage to the virus of existing mainly among men? Let us consider the history of this virus for a possible answer. KSHV-like viruses occur in many vertebrate species, but those most like it are simian viruses. KSHV-like viruses have been found in Asian monkeys, but also in chimpanzees and gorillas that live in the African jungles.

Of all these viruses, those of gorillas and chimpanzees have the greatest resemblance to KSHV. As in so many other cases, it looks like our virus jumped to us from a jungle animal. And again, there is no indication that chimpanzees or gorillas contract KS from the virus; it seems to be completely harmless in those animals.

KSHV must have taken the leap long ago, because man has been carrying herpes viruses for millions of years. It is even possible that a common ancestor of man and the anthropoid apes was infected with an ancestor of KSHV. In that case the human virus has evolved away from the simian virus just as man has evolved away from apes.

This argument is countered to some extent by the fact that people clearly become ill from the virus, whereas apes apparently do not. But with KS in its classical form, people do not become very ill. In addition, illness occurs at a very late age, long after the acute infection with KSHV. The cancer grows extremely slowly in this group of people, who do not have an immune disorder. The form of KS described in South Saharan boys is more aggressive, but it could have something to do with HIV or with some other cause of immune suppression.

KSHV seems to exist mainly in men, and though women can be carriers, they are exempt from the most harmful consequences of the virus. For a virus that remains latent in the body, favoring one sex is apparently one approach to continued virus survival without much injury to the host species. Apparently the compromise between virulence and fitness of the virus consists in this case of the fact that the KSHV does indeed cause cancer, but mainly in that part of the population—men—in which high numbers are less necessary for reproduction and survival of the host. Besides, the men with classic KS are largely past the age of reproduction.

It is not unusual for a virus or other pathogen to pick on one sex more than the other. For example, papilloma viruses cause cancer only in females. They are part of the papova family, along with polyoma viruses. All family members are relatively small viruses with a double-stranded circular DNA as genetic backbone. Papova is an acronym made of *papil*loma viruses, *po*lyoma viruses, and the simian *va*cuolating virus, SV40. The simian virus is one of the polyoma group and is notable for causing vacuoles, or holes, in the cytoplasm that fills all cells. SV40 has given us insight into the group as a whole, but I will discuss mainly the human papilloma and polyoma viruses.

Even more than herpes viruses, all papova viruses are highly specific to their many host species. In fact, they seem to be almost fused with their host. They behave like part of the cell. They barely replicate and evolve at about the pace of their host. We can find their DNA but cannot find a virus particle.

Some of these viruses can scarcely be isolated and grown in vitro. This is a disadvantage for us, because we learn about viruses from their isolation and study. But there is also an advantage, in that these viruses function as a distinguishing characteristic for each host. They are so species-specific that a few unidentified cells can be recognized as cow cells, for example, by the presence of cow papova viruses. The same goes for rabbits and many other mammals, including humans.

Among the human papilloma viruses, the genetic diversity is as extensive as it is among humans themselves. So unidentified human tissue could not only be recognized as human by its viruses, but the human could be recognized as of African, Asian, or Caucasian background. Apparently an ancestor of all mammals was infected with the papova ancestor, and the descendants of both host and virus have evolved together right down to the present. In this respect, it is interesting that the genetic material of these viruses, double-stranded DNA, is the same as that of the host itself.

Some papilloma viruses are the cause of ordinary skin warts. In youth, about 50 percent of all humans are susceptible to these viruses. By age 80, this figure rises to 75 percent. In susceptible people, the skin is crowded with these viruses, though not every one causes an infection or a wart. Just a casual touch can transmit the viruses from one place to another — either on the same host or on a new host. Most skin warts are caused by this virus, a fact discovered only a few decades ago. Before that, theories abounded and superstition blamed "warty" animals like toads.

A completely separate group of papilloma viruses causes warts on the mucous membranes of genital organs. Some of these can cause cancer, specifically cancer of the cervix in women. The papilloma viruses that cause skin warts never cause cancer, and those that prefer the genital area never cause warts outside that area.

Around 1980, a German researcher, Harald zur Hausen, suggested that there could be a connection between papilloma viruses and cancer of the cervix. But there are more than 70 human papilloma variants, and the genital group can itself be divided into two groups. One group carries a serious risk of cancer but the other carries very little risk.

Genital papilloma viruses are transmitted through sexual intercourse, not only in humans but in apes and perhaps other animals. The more sexual contacts, the greater the risk of infection. Approximately one-fifth of all adult women carry papilloma viruses in their uterus. Infections occur in both sexes, especially in teenagers and young adults in their twenties. Every once in a while, such an infection is found in a virgin, but this is very rare. It seems the virus depends for its survival very little or not at all on forms of non-sexual spreading. It infects and is spread by both male and female hosts, but causes serious disease only in females. When the female uterus harbors this virus for a long time, metamorphosis of cervical cells sometimes leads to cancer.

Where did these two papilloma variants come from? The answer will probably never be known because these viruses have been with us—and their other host species—for so long. We can only say they did not come to humans from some other animal. There is no indication at all that these viruses ever jump from species to species, as has happened with viruses like the influenza virus or the AIDS virus. In fact, the most remarkable thing about this group of viruses is how well they hold their own within one single species of host. Within a species and its individuals, these viruses maintain such diversity that they never need to jump.

When a papilloma host must adapt to some unprecedented environmental change, like an ice age, its papilloma viruses adapt too. Whereas some host individuals survive and others do not, however, all of the viruses survive. Those that had the competitive edge under the old conditions do not fall by the wayside. They survive along with those that have the edge under the new conditions. Their continued existence seems guaranteed no matter what happens to the host. In any host, papilloma viruses therefore exist as population of great and ever-growing diversity. It could almost be called a quasi-species, although that term is currently reserved for RNA viruses, in which diversity and adaptation is even more amazing.

Papilloma viruses actually display two kinds of variation: interspecies and intraspecies. In this they are like other viruses—but to a greater degree. The first type allows them to survive in a particular species by means of specific dispersal routes, such as skin contact or sexual intercourse. The other type allows fine-tuning for survival within that species, no matter what happens to the host.

These two kinds of variation are also strikingly evident in the other group that makes up the papova family: polyoma viruses. Besides SV40, this group includes a rodent virus plus two viruses which infect humans, the BK and JC viruses. Each human virus is named with the initials of the person in which it was first isolated. Polyoma viruses are excreted with the urine and infect mainly children. They usually go unnoticed, but in immunosuppressed persons, they have caused brain infections.

Unlike papilloma viruses, BKV and JCV grow easily in vitro. In fact, they grow easily everywhere, and our sewers are full of them (along with enteroviruses, the group to which the poliovirus belongs). BKV and JCV are transmitted from parents to children and among children by way of urine, particularly when toilet facilities are primitive or unclean. The polyoma viruses multiply in cells of the kidneys and urinary tract, so urine is their outlet from the body. Similarly, viruses that multiply in the intestines are spread in feces; viruses that multiply in the lungs are spread through the air; and viruses that produce their offspring in the mouth/throat cavity are spread by saliva.

As is the case with papilloma viruses, humans have been carrying polyoma viruses from time immemorial. And these viruses also do not pass the species barriers. The JC virus is so closely linked to its human host that it has been used to track mass migrations over the past few hundred thousand years. Europe and Africa each have their own JC virus; a large part of Asia, including India, has its own JC virus; China has its own JC virus.

It is a generally accepted theory that *Homo sapiens* originated in Africa less than a million years ago. Small groups of people then migrated out of Africa toward the west, east, and north; gradually the Caucasian and Mongoloid peoples developed in response to the non-African climate and other characteristics of their new homelands.

Polyoma viruses vary according to their population, not the environment. For example, the Indian JC virus is found on the African island of Mauritius, because its population includes people who lived for millennia in northeastern India, then migrated to Mauritius. As might be expected, the first peoples of North America carry an Asian JC virus. They are generally agreed to be descendants of the Mongoloid branch of the human genealogical tree, who crossed from Asia to North America when the islands of the Bering Strait formed a land bridge.

Perhaps the relationship between papova viruses and their host—or better yet, between the virus DNA and its host DNA—is even closer than it is with herpes viruses, since the relationship with its papilloma group eventually leads to cancer in some cases. Do polyoma viruses ever cause cancer? Although the virus is found in cancer tissues from time to time, the question whether it causes cancer remains to be answered. Most scientists believe not, and it seems unlikely from an evolutionary stand-point.

Some light has been shed on this question by study of the simian polyoma virus, SV40. Discovered in 1959 by Maurice Hilleman, this virus suddenly made news when it was found in kidney cell cultures of macaques that were used to make polio vaccines for people. The virus ended up in various vaccine specimens, so many hundred thousand people were ac-cidentally injected with SV40. Researchers hurried to learn more about its effect on non-simian species. They eventually found it can cause can-cer in newborn hamsters—but very large quantities of virus were needed to make this happen. Now, many years later, we have results on all those vaccinated people who unwittingly provided a study group. They have shown no signs of SV40-related cancer, or any other ill effects of note.

If human polyoma viruses ever cause cancer, it is apparently a very rare side-effect of the way these viruses spread, occurring in people with a specific sensitivity to the virus. Most people get the polyoma viruses at quite a young age and carry them all their lives, without ever knowing it.

All these DNA viruses, whether herpes, papilloma or polyoma viruses, set up a close relationship with their host and with their host's cells. They tie their fate to that of their host or, more accurately, to that of the host species. Despite the fact that some spread through sexual intercourse, these viruses are hardly promiscuous. More than most viruses, they are faithful to their host and to their species, never looking for another.

Many DNA viruses co-exist peacefully in the same species and in the same individual. They set up housekeeping in various locations of the body, depending on where the virus enters or leaves during transmission. One person may be infected by several DNA viruses at the same time and seldom notice the effects of any, beyond the occasional cold sore or wart. Thus these DNA viruses survive due to great virus stability per host species and great virus variation within their species. Their closeness to us brings one undeniably negative effect: cancer.

8

RISKING DEATH WITH SEX

The AIDS Virus

An HIV infection leads almost without exception to the death of its victim. Approximately 90 percent of people infected with HIV die from AIDS within 15 years if no drugs are provided. The quantity of virus carried in your blood—the so-called virus load expressed in virus particles per milliliter blood—determines to a large extent how quickly you will die. Someone infected with HIV who, after 15 years, does not have AIDS usually has a virus load of less than two and a half million virus particles. This seems like a lot, but most people with AIDS have fifty to five hundred million virus particles in their blood.

An HIV-infected person's fate is sealed shortly after the infection occurs. Within a year after an HIV infection has been contracted, the measurable quantity of virus in the blood becomes stabilized at a level that predicts how you will end up. As a rule, the more virus at this crucial point, the more chance that AIDS will develop. Some people carry such a large quantity that they die within a year after the virus has penetrated their T cells. In most cases, the immune system can offer resistance for about 10 years, but then your time is up.

The few lucky people with low or undetectable levels of virus in their blood one year after being infected have the best chance—unless the levels jump unexpectedly over time. We call them "slow progressors." They too are headed for AIDS, but so slowly that they may live long enough to die of some totally unrelated cause.

It seems that a little bit of HIV in your blood can't do much harm. In fact, after HIV was discovered in 1983, it took until the mid-1990s for us to grasp why people get AIDS. It turned out that the more HIV in the blood, the faster the immune system breaks down, and the faster all kinds of other viruses, bacteria, and parasites are given free play. AIDS is in fact a hodgepodge of illnesses that are caused by all kinds of pathogens. In fact, HIV is a unique virus in that, by itself, it does not cause the disease symptoms that eventually occur with HIV infection. Instead, it causes a complete breakdown of our immune system, which guards against attacks from the outside as well as the inside.

Viewed very objectively, HIV shows what we are made of. In every person it infects, it reveals what the immune system can stop and what it cannot. Like our faces, our immune systems—and everything else in our bodies and minds—are basically similar, but with wide variation. Just as some people have a better musical ear, some have more effective immune systems. As with a musical ear, you can improve it with training, but only to a point.

People with HIV contract illnesses that they would never normally contract—or which are much worse than they would normally be. For example, AIDS was first recognized in the early 1980s because of a strange epidemic of pneumonia. It was caused by *Pneumocystis carinii* in homosexual men whose blood turned out to have a surprisingly low number of T cells. Anyone with a normal amount of these immune cells would never get *P. carinii* pneumonia (PCP), even though this parasite is all around us. The HIV infection, which had depleted the T cells, was the only reason why homosexual men in New York and San Francisco were suddenly dying from this "opportunistic" infection. HIV had provided the opportunity for an innocent parasite to become virulent.

The T cells are named for the thymus, a small gland where they are produced. Of the many types of immune cells, they are among the most important. Unfortunately for us, they are HIV's favorite cell if they have certain features, chiefly a receptor molecule we call CD4. All cells positive for this receptor (CD4$^+$ T cells, or simply "CD4 cells") attract HIV and allow it to enter. It then reproduces, and the T cells burst apart to release the new generation. Too bad for us that HIV likes our T cells instead of the more expendable types of cells that satisfy most viruses.

Even before HIV does enough T cell damage to cause AIDS and opportunistic infections like PCP, it causes problems. In ways that are unclear, it makes us more vulnerable to such diseases as tuberculosis. It favors certain disease manifestations, such as blindness due to cytomegalovirus infection or Kaposi's sarcoma due to infection with the herpes virus HHV8. Tuberculosis had been a problem in Africa for a very long time before AIDS reared its ugly head, and Kaposi's sarcoma (KS) also had a long history. It was first diagnosed in an African man by Viennese dermatologist Moritz Kaposi in the 19th century. But what he saw was minor compared to the KS now seen in HIV-infected patients. All the above infections have a more serious outcome in the context of HIV infection. Or they are contracted at an earlier age than in non-infected people. Here again, the underlying cause is a breakdown of the immune system due to HIV.

However, there is a unique group of people who are more or less immune to HIV, or whose immune system is scarcely affected by an HIV infection. They are not the slow progressors who become infected but get AIDS very slowly. They are, for example, prostitutes in Kenya who do not get infected despite daily sexual relations with infected men. They are homosexual men with a very active sex life who somehow manage to resist HIV for decades. How is this possible?

We now know that their cells are missing a part that HIV needs in order to penetrate. It cannot get into the cells and therefore cannot cause infection. During the years 1983–1984, studies were conducted on HIV-infected people with AIDS or at risk to develop AIDS. The virus was isolated from T cells in their lymph nodes and from T cells circulating in blood. The big question was: If HIV causes AIDS by damaging the immune cells, how does it get in? The studies found that when T cells with CD4 on their surface were blocked at that receptor, HIV could not enter. So CD4 was the answer—at least, it seemed that way.

The next question was: Do the homosexual men and Kenyan prostitutes who remain uninfected despite constant contact with HIV have no CD4 molecules on their T cells? Or not enough? Or are the molecules blocked in some way? To the amazement of everyone, their CD4 molecules were no different in number or configuration than those of the people who got infected.

It looked like there had to be a co-receptor, a second molecule on the surface of the $CD4^+$ cell that cooperated to give entry to HIV. About 10 years later, in 1996, the co-receptor was found: CCR5. It's just as important as CD4, which retains the title "receptor" only because it was discovered before the "co-receptor." (The same applies to "CD4$^+$ cells," which could just as well be called "CCR5$^+$ cells." Actually, CCR5 is one of several in a group of so-called chemokine receptors. But it is by far the most important, letting in the vast majority of HIV strains that seek entry. A small number of HIV strains use another molecule, CXCR4.

Recognition of the role of CCR5 emerged from the finding that it was totally impossible to infect certain men with HIV, and that these men, while having CD4 on their cell walls, did not have any CCR5. The reason turned out to be an inherited deficiency in their CCR5 gene. Genes are often named for their product: the protein molecule for which they carry the code, or recipe. Genes are made up of thousands of nucleotides, sometimes more, and in these men, the CCR5 gene lacked just 32 nucleotides. But this deficiency was large enough that no CCR5 protein whatsoever could be made.

Actually, since we have two copies of every gene in our chromosomes, there are three kinds of people with respect to CCR5: people with two normal CCR5 genes, people with one normal and one abnormal gene, and people with two abnormal genes. Those with two normal genes —the vast majority—will easily contract HIV infection. Those with one abnormal gene and one normal gene will also be infected, but they will less quickly get AIDS. And people with two abnormal genes will never be infected with HIV. For them there is only the minuscule risk that they will become infected with the rare HIV strain that uses the CXCR4 receptor.

We might call the abnormal CCR5 gene the "AIDS resistance gene." Like many mutations, it is found more in some populations than in others, because different groups have different histories. Who has the AIDS resistance gene? It is truly remarkable that not one African or Asian person has been found to have it. It has been specifically searched for and ruled out in China and Korea. It has been ruled out among Arabs in the Middle East. However, Jews in Israel, Iraq and Iran have a high incidence of this gene. It occurs most often, more than anywhere else in the world, among the Ashkenazi Jews of Eastern Europe and, somewhat less often, among the Sephardic Jews of Southern Europe. Ashkenazi Jews are therefore more resistant to HIV than any other population group on earth.

We do not know exactly why this is, but as with KS (see chapter 7) it somehow stems from Jewish history and migration. As explained before, when the Romans drove the Jews from their homeland, about 2000 years ago, they went in two main directions. What we now call the Ashkenazi Jews went north through Asia Minor into Europe, with some remaining in Asia Minor. The Sephardic Jews spread west along the North African coastline and up into Spain. All indications are that the AIDS resistance gene originated in northeastern Europe, where it is still found among both Jewish and non-Jewish inhabitants of Sweden, Finland, the Baltic states, Russia and Poland.

Much of northeastern Europe was once known as the Pale of Settlement, a region stretching from Estonia, Latvia, and Lithuania through White Russia to Ukraine. It was a fringe area to which Europeans tried for many centuries to confine the influx of Jews. It was in the middle of nowhere, and today, if we say something is "beyond the pale," it is geographically or socially "far out."

At some time in the distant past, an inhabitant of the Pale must have been born with an accidental mutation in the CCR5 gene—the AIDS resistance gene. It seems to have originated between the early Middle Ages and the end of the 17th or the beginning of the 18th century. However, instead of "originated," an evolutionist would say "was selected out," because here is what happened. The person with the mutation mated and produced others with the mutation, eventually resulting in a small group. A group with such a gene tends to continually grow—but not because people with this gene are more fertile or have more children. They are simply "selected out" because people without the gene die earlier, and before they have been able to replicate themselves by having children.

Perhaps people with the AIDS resistance gene had a special advantage even before AIDS came along. It may have protected against another infectious disease which, during the seventeenth and eighteenth centuries, precipitated the death of millions of people who lacked the gene. This could have been the plague, but it probably was smallpox. In fact, we have recently learned that some pox viruses enter their favorite cell with the aid of the same surface molecules that let HIV into T cells (see next chapter). Perhaps the AIDS resistance gene was originally a "smallpox resistance gene" that just happens to afford resistance against HIV infection. This would of course be quite a coincidence.

A more straightforward course of events would be that, a few hundred years ago, the AIDS virus itself caused an epidemic in northeastern Europe that decimated people lacking the AIDS resistance. But there is no indication that this happened. Everyone now agrees that the AIDS epidemic dates only from the twentieth century and that HIV is a relatively new human pathogen.

Major questions remain as to exactly when and how HIV emerged and how a few isolated cases became an epidemic. One theory which caused quite a stir was proposed with fervor by E. Hooper in his 1999 book, *The River: A Journey to the Source of HIV and AIDS*. His idea is that HIV was transmitted from chimpanzees to people at the end of the 1950s through polio vaccinations in Africa. The polio vaccine in question is a live poliovirus as opposed to a killed virus. But it is a weak poliovirus that causes no illness and provides protection against the true poliovirus. According to Hooper, this polio vaccine was made by cultivating the weak polio virus in tissue taken from chimpanzee kidney, and the chimpanzee cells were already infected with the AIDS virus.

This polio vaccine was used between 1958 and 1960 to vaccinate inhabitants of Leopoldville in Congo against polio. It is Hooper's supposition that the chimpanzee virus was spread among people in this manner. But recent studies have shown that the kidney cells used to prepare the polio vaccine came from rhesus monkeys and guenons, not chimpanzees. In addition, no trace of the AIDS virus has been found in stored samples of vaccine preserved since that time.

All indications are that the AIDS virus was spread not by large groups of people being injected with a chimpanzee virus but by a few isolated cases of exportation. Viruses that had for a long time been circulating in the interior of Africa among a small group of the local people came out in the open. This happened in various ways, and only later were the outbreaks linked to HIV. For example, in the 1990s, an epidemic occurred in Romanian orphanages. It was later traced to blood from an infected donor, who turned out to be from the Congo.

Before it could spread in Europe, HIV had to spread in Africa. How would this have happened, especially in those days of sparse population, isolated villages, difficult travel, and conservative sexual practices? My own theory, described in *Viral Sex: The Nature of AIDS* (Oxford, 1997), is that a few infected Africans were among the many that colonial powers drafted as soldiers to fight alongside their own troops. The resulting rag-tag armies ranged over wide areas to put down local revolts or to struggle with rival colonial powers. It appears that HIV arose in Cameroon in Western Africa and was then carried, about 1900, to the area around Lake Victoria in Eastern Africa. Germany then controlled that area but lost it after World War I, and many German colonists returned to Europe. In my previous book, I describe a 1939 epidemic of pneumonia due to *Pneumocystis carinii*, an infection which points to an underlying immune disorder. This epidemic took place in Danzig and originated with farmers who returned to that port city from German colonies in Africa.

If HIV was carried from West Africa to East Africa, it might well have caused a small epidemic in each area, unnoticed at the time. In remote circumstances, the absence of records does not mean outbreaks did not occur. And recently, studies not influenced by my own or by each other seem to support such events. In analysis of the genetic history and variation of HIV, Bette Korber from the Los Alamos National Laboratory and Anne-Mieke Vandamme from the University of Leuven showed, independently, that a small group of people must have been infected with HIV sometime between 1900 and 1945. In addition, data from Francine McCutchan's group in Washington, D.C., point to two early AIDS epidemics in Africa—a West African and an East African epidemic—each with its own characteristic group of strains.

How the very first human in Africa became infected with HIV remains a puzzle, but many of its pieces have now been put in place. HIV is probably the human version of a virus that jumped from an animal. This animal would have been enough like us that the virus could gain a foothold and adapt to the new environment. Of all animals, we are closest to our fellow primates, which are divided into simians—the apes and monkeys—and the prosimians, like lemurs (which do not concern us here). Apes are generally larger than monkeys and, unlike most monkeys, they have no tail. Sometimes called "anthropoid apes," they include gorillas and chimpanzees, among others. Among monkeys are such species as baboons, guenons, mangabeys, vervets, and green monkeys.

There are simians around the world, but our focus is Africa, where apes and monkeys are infected by a virus very similar to HIV. By analogy with HIV, it is called the simian immunodeficiency virus (SIV), even though it rarely harms these non-human primates. In general, each species of ape or monkey is infected only by a certain SIV population, though within that population there is strain variation. (They are alike but variable, like the people at a family reunion.) The virus population has evolved with its host and is therefore named for it, e.g., SIVch for chimpanzee SIV.

So there are countless SIV strains in Africa, and one could surely have jumped to a human and become HIV. But finding that one has been much harder than anyone expected. And it must have been an unusual strain because we find few SIVs that are close enough to HIV to have made the transition. In fact, after exhaustive study, we have found only one SIV that closely resembles HIV. It is a very rare SIVch. After its discovery, thousands of chimpanzees were tested by Beatrice Hahn's group in Alabama and by Walid Heneine's group in Atlanta. Only a very few of these chimpanzees carried HIV-like viruses. More recent studies found a significantly larger proportion of chimpanzees infected with SIVch, making the chimpanzee-origin of HIV more likely.

In 1999, Hahn described in *Nature* another rare strain, but this time it was HIV. It occurs only in Cameroon and includes genetic information that came from a chimpanzee virus by recombination. This finding convinced many that the HIV now causing our worldwide epidemic came originally from a chimpanzee. They surmised that transmission occurred when an African ate the meat of a chimpanzee, a not unusual circumstance.

However, there are three problems with the 1999 report. First, the genetic analysis was a protein analysis, which is considered less reliable than a DNA analysis. Second, relatively few chimpanzees in the wild are infected with any chSIV that could have been the recombinant "parent." Third, the HIV that had absorbed genetic information from a chimpanzee virus has never spread outside of Western Africa. Nor is it a close relative of the HIV now causing the worldwide epidemic. It is only a distant cousin.

Already in the early 1990s, some researchers who accepted the simian origin of HIV had expressed serious doubts about a direct transmission from chimpanzees to humans. They began to think the fateful virus originated in a smaller simian species from which it was transmitted to both chimpanzees and to humans. It would be smaller than chimpanzees because, otherwise, how could a chimpanzee eat it? It would be a monkey, but the monkey with just the right characteristics has not yet been found. In fact, though African monkeys are widely infected by SIV, their SIVs are less like HIV-1 (the type causing the worldwide epidemic) than are the SIVs of apes. Monkey SIVs are more like HIV-2. This virus is far less notorious than HIV-1 because it less lethal and has hardly spread from its original African habitat. The evidence for HIV-2 originating from a monkey virus is much stronger than for HIV-1; the major species jump appears to have occurred in the 1940s, later than the one estimated for HIV-1. (In this chapter, HIV refers to both types or only to HIV-1.)

As noted earlier, African apes and monkeys generally have no trouble with their SIVs. The SIV of one species might sicken another species, and all SIVs can sicken Asian monkeys, who have no SIVs of their own. But African apes and monkeys live happily with these viruses even though

SIV, like HIV, infects immune cells and sometimes enters by the very same types of molecules. Most of the individuals in any African simian population are SIV-infected early in life. Among some species, like guenons, there is hardly one mature animal that is not infected.

Most astonishing, the SIV load in monkeys (though not in apes) is often as high as the HIV load in a person with AIDS, yet the monkeys stay healthy. Their immune system remains intact. Probably monkeys on this earth once suffered as much with SIV as we do with HIV, but some individuals of the species had an "SIV resistance gene" and/or could live happily with a high virus load. They were selected out while the others gradually died off. So monkey SIVs now survive comfortably in a host that is not sick and, in addition, is no doubt protected by its SIV from more aggressive variants. Everybody is a winner.

However, the fact that African apes and monkeys are not bothered by SIV or HIV is a mixed blessing for us. We are happy for them but badly need a model in which we can study HIV and AIDS. At first, progress with HIV was especially slow because we had no model. It seemed that only primates were infected by SIV or HIV, and no non-human primates were made sick by either one.

Finally we learned that Asian monkeys can be made sick by SIV, and crucial research is now ongoing using rhesus macaques. Then we learned, just a couple of years ago, that aggressive variants of SIV or HIV exist or can be evolved in the laboratory. Such variants can cause illness even in animal species that exhibit the greatest resistance against the ordinary variants. With only a few mutations in the genetic material—spontaneous or engineered—a virus can turn into a monster able to break down the resistance of its usual host.

In one famous case, a chimpanzee was HIV-infected in the laboratory and, unlike all previous cases, it finally developed AIDS. It was the only case of illness to occur after hundreds of chimpanzees were experimentally infected with HIV during the early years of the AIDS epidemic. AIDS took 10 years to develop in this animal, but when its virus was isolated and injected into another chimpanzee, the second animal became sick much more quickly. This proved that first ape was not unusual but that the virus was unusual. Moreover, during its infection of the first animal it had become more able to cause illness in chimpanzees due to changes in its genetic code. It lowered the number of immune cells in its animal hosts and also produced an enormous quantity of virus in the blood.

An AIDS-causing virus does its damage because it strives in every host to reach very high levels. It does this not to make us sick but to assure its own transmission and survival. However, its high production kills so many immune cells that opportunistic illness occurs as a side effect. The more HIV a person has in his or her blood, the more easily and quickly the virus can spread. Selection therefore has taken place, leaving us to battle with HIVs that are highly productive. With other types of viruses, high levels cause less harm, because they do not infect immune cells. And even SIVs, which infect simian immune cells, do not harm their usual simian hosts. But humans cannot cope with so many HIV virus particles in their blood, and AIDS develops. There is thus a direct relationship between illness of the host and spreading of the virus.

It has generally been assumed that a parasite never wants a sick or dying host—that a virus is better off if its host does not become too ill. However, the assumption seems doubtful here. HIV actually has trouble dealing with its human host and must overproduce to keep going. In a population of HIV-infected persons receiving no therapy, the average individual virus load is 10 to 50 million virus particles. That amount of virus will cause illness within 5 to 10 years.

Since HIV is transmitted sexually, infection occurs especially at an age when sexual activity is at its peak. But since the external characteristics of AIDS tend to curb sexuality, with declining energy and physical appeal, transmission generally takes place early in the infection. After the brief acute phase, during which people feel flu-like symptoms, they feel perfectly healthy for months or years. This period apparently allows the virus enough time to spread and survive.

Why does HIV need to overproduce? Because, contrary to its reputation, HIV is much harder to spread than many other pathogens. Its transmission demands the exchange of blood or certain other bodily fluids that carry the infected immune cells. It therefore demands the most intimate direct contact between donor and recipient, spreading mainly between sexual partners or between a mother and child during the birth process or breast-feeding. Once transmitted, HIV must reach the circulating blood to find and infect its chosen cell in the new host.

In contrast, most other viruses spread in the air, on contaminated objects, or by casual contact between people. A flu virus spreads with a cough or sneeze to anyone within several feet of an infected person. It wafts effortlessly through the air and, when inhaled, comes immediately in contact with the cells that have the receptor for the flu virus. They are right there in the airways.

The good news about HIV is that, since 1996, there is a very effective treatment for HIV infection: highly active anti-retroviral therapy, or HAART. This prevents the development of AIDS or postpones it for a very long time. It is complicated, expensive and has many side effects, but it definitely lowers the quantity of virus in the blood by 99 percent. Large-scale use of HAART would not only prevent AIDS in those being treated, but it would also lower the number of new infections, because HIV-infected people would be less contagious, having much less HIV in their bodily fluids.

HAART has been of enormous benefit to individuals. It should also have had an enormous effect on the transmission of viruses but, in reality, its effect on the spread of HIV is quite disappointing. The main reason is that only 300,000 people worldwide are getting this treatment, while 30,000,000 people are infected with the AIDS virus. So only 1 percent of those infected with HIV are treated effectively and, almost without exception, these people live in the most affluent parts of the world: western Europe and North America. Meanwhile, 15,000 new HIV infections occur every day, especially in Africa and Asia.

In the long run, we need a vaccine. We must combat AIDS by protecting the people who have not been HIV-infected rather than by reducing the illness and contagiousness of those already carrying the virus. The vaccine should be as cheap and easy to use as we can make it.

The first generation of HIV vaccines is already being tested on humans. It is not based on HIV itself but on other viruses, including pox (see next chapter) and relatively harmless adenoviruses, which have been engineered to protect against HIV. Projected for clinical use in about 2010, it is designed to keep AIDS from developing by sharply reducing the virus load. Especially if given early in the infection, it helps the immune system check the virus by generating more HIV-specific immune cells.

These cells are a kind of T cell that fight HIV-infected cells, and they work very well for awhile as is learned from monkey studies. Since HIV is allowed to infect the cells, it will slowly but surely multiply even in the presence of many immune cells. Sooner or later, with much individual variation, it might eventually outnumber or escape from these defenders, and then AIDS would re-emerge.

This vaccine is undeniably a big step forward. It would certainly extend lives—but its effect could well be limited and of relatively short duration. After all, it just shifts a lot more people into the group that naturally have a low quantity of virus in their blood: the slow progressors. And the speed of this shift, within any given group of infected people, depends on three main variables. The vaccination campaign must be efficient; the resulting drop in virus load must be adequate; and the vaccinated persons must do their part.

This last factor is of supreme importance and of great concern to everyone working on HIV and AIDS. Already, studies of people on HAART have shown that effective treatment causes many HIV-infected people to increase their sexual activity and risky behaviors. First of all, they feel better and look better. Second, they know that their quantity of virus has decreased, so they feel less infectious to others. Unfortunately, they can still pass on their HIV infection, if only at a much slower rate. Like people on HAART, vaccinated people might be inclined to increase their sexual activity and their danger to other people.

It is very likely that in some individuals the virus will escape the cytotoxic T cells, as mentioned above and such a vaccine-resistant virus might spread from host to host. When that happens, these vaccines have lost their purpose. At best, the first generation vaccine may have a very limited effect on the AIDS epidemic and should perhaps be seen as a stopgap: something we use only until we can develop a vaccine that blocks HIV infection entirely. However, this will take many years and may not even be feasible.

Let's be very realistic: we have come a very long way since 1983. We are now poised to keep at least one step ahead of the virus—and that may be all we can expect or need. Can we truly expect to completely stop the spread of a virus which, for its survival, is so dependent on that one activity—sexual relations—that also guarantees the survival of human beings? Nature has made this activity very attractive, and sex is not our only problem. We are facing a retrovirus that can splice itself into the nuclear material of the infected cell and survive as long as the infected cells survive. It will disappear from the body only when all the infected cells have disappeared from the body. No treatment thus far can accomplish this, nor can any vaccine, so perhaps AIDS will always be with us in some form.

In the case of smallpox, the illness was eradicated together with the virus. In the case of polio, we thought we had done this, but already there are doubts. In the case of HIV, we may have to be satisfied with stopping AIDS without stopping the virus. We can proceed in two ways. One, we could treat every HIV-infected person with antiviral medications before AIDS manifests itself. Two, we could vaccinate all uninfected persons in the world with the goal of allowing HIV infection but drastically lowering the subsequent virus load.

Neither approach aims at complete eradication but, even so, they present a huge challenge. Their success depends on our ability to reach millions of people with drugs and vaccines, in the case of drugs even for a lifetime. With tens of thousands of new HIV infections per day occurring in China, India and Africa—often in very inhospitable areas—this might well be too much to ask. Billions of dollars would be needed, and the prosperity and infrastructure of developing areas would have to come much closer to ours in the Western world. We can only hope that this somehow happens in time to benefit all the HIV-infected people in the world and the mothers, fathers and children who are exposed to the AIDS virus every day.

Otherwise, the only solution may be a live HIV vaccine, despite the inherent danger. After years of research, such a vaccine is increasingly becoming a feasible alternative. It would require major funds and effort, however, but much less than the first-generation vaccine for this reason: given a push it would maybe even spread itself. Ideally it would not have to be injected into each and every person in the world. A weak HIV strain would be helped to establish itself and block the true HIV, from which we would then be protected. This is what we did in the early days of smallpox vaccination: cowpox edged out the stronger virus. Since then, we have used this approach successfully with diseases such as polio and measles.

Ideally, a live but weak strain used in vaccination is unable to cause much damage. The ideal HIV vaccine would therefore never be able to cause AIDS. Or it would cause AIDS only at a much lower frequency than true HIV and much later during the course of the infection. Already various researchers, including Ron Desrosiers in Boston, have developed a weak SIV that protects rhesus macaques against a true SIV. The true SIV makes them ill because they are Asian monkeys with no SIV of their own and no natural resistance to SIVs. The weak stand-in is so weak it normally cannot infect even these monkeys, but with human help it can. Desrosiers has shown conclusively that it prevents not only disease but any initial infection by the true SIV. So far, this vaccine is the only one that can do this, and it represents a big step toward an HIV vaccine of the same type. If we had such a vaccine for HIV, the true HIV would be entirely kept out of the host cells. It could not hide there and multiply, even slowly, causing trouble down the road. It certainly could not cause disease.

So what is the catch? What is the inherent danger that was mentioned above?

We know a live, weak SIV can compete successfully with the true SIV, and we know how this works. Like HIV, SIV is a retrovirus, and this family of viruses is especially known for a phenomenon called *superinfection interference*. Once a retrovirus has penetrated a cell, it denies any later-arriving virus access to that cell. Once someone has been infected with a meek and mild AIDS virus, it is there to stay, no matter how many vicious AIDS viruses try to penetrate cells in the body.

The key word here is "stay."

Will such a vaccine virus stay forever meek and mild? We do not know, which is why all researchers are not totally enthusiastic about a live, weak HIV vaccine. How weak is weak? How—and when—would we know when we have the right level of weakness? The Desrosiers SIV vaccine protects normal adult monkeys but eventually causes AIDS in susceptible animals, such as newborns. This is a major worry since much HIV is transmitted vertically, from mothers to babies. Clearly Desrosiers' strain needs to be further weakened—but without making it too weak to compete with the true SIV.

Let's assume we can create an SIV or HIV vaccine of just the right strength for widespread use and maybe even an HIV vaccine strain that can be switched on and off by drugs. Will it stay that way? What if it then changes over time and we have created a monster? For better or worse, it will stay in the body of everyone who is vaccinated. If we decide later that we don't want it, we cannot get rid of it—though there might be ways we could modify it.

Of course, researchers like Desrosiers are working furiously to know the weak vaccine viruses better. Already it looks as if the infection caused by an HIV vaccine strain would need about a year to establish itself. Until then it would not provide full protection against more aggressive variants. Also, it looks as if a vaccine strain could potentially generate a new aggressive virus in the laboratory. This has occurred in a test tube, by means of mutations, and could perhaps occur in nature. A more aggressive virus could also develop by means of recombination, if the weak vaccine strain mixed genes with those of a related virus.

So probably we can never be sure that a live, weak HIV vaccine will remain weak and safe. Sooner or later it could cause AIDS with a certain regularity, though much less often than with the true AIDS virus. The approach would work—it would displace the virus we have now. But mass vaccination with a live HIV vaccine, which has a 5 percent risk of causing AIDS over 25 years, has been compared in mathematical models to the true AIDS virus, with its 95 percent risk over 25 years. The prediction is that, over 50 years, the vaccination would lead to much less AIDS in a country that now has high AIDS mortality. However, over the same period, it would lead to more AIDS in a country that now has low mortality.

In either type of country, the weakened HIV vaccine would eradicate the true HIV by replacing it, over time, with the vaccine strain. As long as vaccination continues, the vaccine virus will remain endemic, or embedded in the population. When a pathogen is endemic (Greek, "within the people"), it is part of the landscape but may be invisible except for occasional outbreaks or "epidemics," when it comes to the surface of the population. As was the case for smallpox, the vaccine strain can itself be eradicated when it has totally replaced the true virus. The only danger is that the vaccine strain by that time has changed to an aggressive virus that can keep itself alive. Theoretically, such a change is very possible indeed.

The only hard data about the possible usefulness of a live AIDS virus come from a recent epidemic among horses in China. It was caused by a horse virus, the equine infectious anemia virus (EIAV), a virus in the same family as SIV and HIV. It first appeared in the offspring of Russian horses purchased by the Chinese as breeding stock. They wanted larger horses for use in guarding the northern border, so large Russian horses were acquired to mate with the smaller horses native to China. Together they were expected to yield horses of the desired size—but the first generation had the new disease.

The research center for animal medicine in Harbin, in northern China, was commissioned to find a remedy against this equine tragedy as fast as possible. Chinese researchers developed and used a live EIAV vaccine in the 1970s and 1980s to eradicate EIAV. The weakened live vaccine virus prevents three out of every four infections with the true EIAV. On the other hand, the vaccine itself causes a fever in one out of every four vaccinated horses; anemia is the underlying cause (*equine infectious anemia*, but never causes death in the vaccinated horses. Nor has the vaccine virus ever been seen to spread on its own.

In other words, this live EIAV vaccine affords 75 percent protection, with an illness frequency of 25 percent as a result of the vaccine strain. The illness caused by the vaccine strain struck two or three months after vaccination. The illness caused by the true EIAV produced fever within a few weeks and resulted in death for most of its victims. So the vaccine was a big improvement. It was injected into hundreds of thousands of horses, halting the EIAV epidemic in China and apparently sweeping the illness from Northern China. That is, the true EIAV is not endemic and will most likely not break out again.

Would it be possible to make such a vaccine for HIV—and would it be accepted by the world? The development of such a live vaccine was advocated by Howard Temin in one of his last articles before he died in 1994. Together with David Baltimore, Temin received the Nobel Prize in 1975 for the discovery of reverse transcriptase (RT)—the enzyme that lets retroviruses convert their RNA to DNA.

As was detailed in chapter 6, Temin had noticed that all animal species except humans are regularly subjected to attacks by simple retroviruses, which have only three genes. In the laboratory, these viruses can thrive in human cells, but in nature they cannot enter and infect our cells. We know that simple retroviruses plagued our long-ago ancestors, because we find remnants of their viral DNA in the DNA of ancient mummified humans. But somehow humans were eventually able to knock out simple retroviruses for good (see chapter 6).

The question is whether we can do this now with complex retroviruses, such as HIV, and whether we can do it faster than it would occur in nature. Temin advocated this strategy in light of the rampant AIDS epidemic. Whereas HIV normally has nine genes, he suggested making an HIV stand-in with only three genes. This dummy virus would be able to reproduce effectively in human cells but would be unable to do much damage. It would block the true HIV.

In 1993, it seemed impossible to develop in the laboratory a virus that would spread effectively in human populations without, at the same time, causing AIDS. But recently a bright light has appeared on the horizon. In an experiment, an HIV virus was weakened due to mutations in the NEF gene, among others. The resulting virus could barely reproduce but was assisted by inducing another mutation to compensate for the adverse effect of the NEF mutation. Then the virus was able to reproduce rapidly—but it retained the crippled NEF gene. This gene is thought to play a crucial role in the process by which HIV damages our immune cells. The compensatory mutation affected a part of the virus called LTR—a part that complex retroviruses have in common with the simple retroviruses. So it has now become possible, in theory, to make an HIV strain in test tubes that fills our two main requirements. It breeds up to a large virus load in humans and can thus spread effectively, but it cannot cause AIDS because it cannot damage the immune cells.

As the cowpox virus was used against smallpox, such an HIV could serve as a vaccine against the HIV that causes AIDS. It would be able to out-compete the true HIV and eliminate it from the world. AIDS would be stopped and become part of the past. A problem solved. In centuries to come, it would be almost forgotten, an illness of the 20th and 21st centuries that we have put behind us.

Temin's idea may not be impossible after all.

WARRING AGAINST HUMANS AND OTHER ANIMALS

Smallpox, Monkeypox, and Others

In today's world, there are no more natural cases of smallpox. The last occurence was recorded in Somalia, October 26, 1979; the victim was a cook who worked in a hospital. But smallpox did not disappear from the face of the earth by a natural accident. Smallpox as a disease has disappeared because of a systematic worldwide effort to cut the virus off from all means of surviving in the wild. The smallpox virus is widely considered to be eradicated, even extinct. But it is still present in freezers around the globe: the United States has virus stock, and Russia and even some terrorist groups may have the virus.

In nature, the evolution of living things leads inevitably to extinction, though the process may take millions of years. Most scientists now divide living things into three main groups: *archaea* and *bacteria*, for all the one-celled organisms, and *eukarya* for all the rest. The eukaryotes, with many cells and cell-types, include all plants and animals, from weeds and worms to humans and trees. The three life forms see the extinction of some of their vast numbers of species every day. Over the millennia, many species have become extinct, and others are on the way to extinction at this very moment.

Generally a species dies out when it is unable to adapt to gradual or sudden changes in its natural environment, such as extreme cold or extreme heat. Archaea are perhaps the least vulnerable. As indicated by their name (*archaeon* in the singular), they are probably the world's oldest life form. They seem to resist and even enjoy conditions that seem to us extreme and unbearable. Nicknamed "extremophiles," they have surely lost some species to extinction, but some of those we see today have slogged through the earliest and hardest times on our planet. They have survived all the ice ages and boiling upheavals that have come along since life first appeared on earth. Like *sulfolobus*, which thrives in hot sulfur springs, they nestle in the most inhospitable corners of the world. Sulfolobus can reproduce in an environment of extreme acidity with a pH of 1 to 3, at temperatures as high as 90 degrees C.

Where are viruses in this vast and ancient scheme of things? They are everywhere, even in those hot sulfur springs. Like sulfolobus, all its parasites—including viruses—have managed to resist the most extreme and potentially lethal conditions. Most life forms support parasites like viruses. Viral parasites are not themselves considered fully alive, mainly because they cannot reproduce without the help of their host. Viruses get that help by entering a specific cell in a specific host, then subverting its reproductive machinery to produce their own offspring.

A few viruses can go even further in their sabotage. Their genetic material is RNA, not the DNA of all living things (and some viruses). But with a special enzyme, these super saboteurs can convert their RNA to DNA and splice all or part of their genome into the DNA of their host cell. When even a scrap of viral DNA becomes part of a host's DNA, it is permanently integrated and will be inherited when the host multiplies. It will survive until all species members who harbor that piece of viral DNA become extinct. And if the host is an archaean, it can take quite a beating.

However, only retroviruses can integrate. Most viruses are like the smallpox virus, which infected humans for millennia but never became endogenous, or inborn. Such an insider may become extinct as soon as its host becomes extinct. In contrast, an exogenous virus could theoretically be driven to extinction even if its host continues to live. The question that especially intrigues scientists is whether any *human* viruses have ever become extinct. Smallpox is the one example that springs to mind—but is it truly extinct? How would such extinction occur and what would it really mean?

As we know, a virus must infect a certain cell in order to reproduce. The virus cannot think about this at all, or even look for the right cell. Infection just happens when a virus is chemically attracted to a certain cell, becomes attached to certain receptors or entry molecules, and wiggles inside. The virus then waits to see if the cell enzymes will multiply its genetic material. If so, the host cell will soon give birth to the next generation of viruses (also called virions or virus particles).

Sometimes the reproduction and birth process requires so much effort and energy from the host cell that it will sicken and even die. Often the cell must burst apart to release the offspring. Or the particles will pop through the cell wall, one by one, taking bits of wall with them as their own envelope. Such damage to host cells may or may not hurt the host in a major or permanent way, but in any case, the virus survives. The new particles are already on their way to infect a new victim and produce the next generation.

Of course, a living organism does not usually lie there and let viruses propagate undisturbed. Most organisms have an immune system that attempts as quickly as possible to wipe out enough of the invader to stop infection or reproduction. The virus survives only if it can keep ahead of the immune reaction and get its offspring on their way to a new victim before the host kills them. So viruses always try to run ahead of their host's immune system. They develop all kinds of tricks in order to escape just in time from the immune fighter cells. Viruses are so good at this that it probably only rarely happens that they become extinct without their host becoming extinct.

Probably a virus can die out only if it depends on just one species, and that species stops reproducing the virus. Besides host extinction, one way this can happen is that the species—let's say man—undergoes a genetic modification that eliminates the cell needed by the virus, or makes that cell less friendly. The virus cannot find its favorite cell, or it finds the cell but is blocked at one of three crucial points: cell entry, reproduction inside the cell, and release of offspring from the cell.

A second scenario is that some virus has already moved into our hero's favorite cell. This interloper is usually a closely related virus, because otherwise it could not use the same receptor. In any case, the cell is already infected and cannot usually be super-infected. And the interloper has stimulated the host's immune response, so the intruding virus is blown away.

This second scenario, based partly on the phenomenon of "superinfection interference," partly on activating the host immune system, is the basis for vaccination. We block one virus by filling its home cell with a related virus—one that is much weaker and does us less damage.

So can we say that human smallpox is extinct? It was a human virus, dependent only on us, and no cases have appeared for almost 30 years. Before going further, we must agree on our definition of a *human virus*. Obviously it is one that can infect human beings. (Infection may or may not make us sick, though this is what people remember about viruses.)

Must it infect only human beings? Most viruses can infect only one species, a very few species, or a family of related species. However, many of what we now consider human viruses originated in other animals and can still infect those animals. These viruses stumbled on humans by accident or when their usual host became scarce. Over time they converted to a human virus, infecting us exclusively while their relatives, back in the animal world, went on infecting their favorite beast.

These converts to humans could, in some cases, jump back to one of their "animal pool" if humans become scarce, but some could not. For example, certain hog and duck viruses jumped to humans in China, where farmers live closely with their animals. These viruses are now so adapted to humans that they can no longer infect hogs or ducks. They could therefore be considered true or exclusively *human viruses*: not only can they infect humans, but they can infect no other species. They would die out before they could evolve enough to thrive in a new host.

So the smallpox virus is a true human virus because it can infect us and has no alternative host. But if it is now extinct, its one host clearly is not—though human extinction is not impossible. Apes, our closest primate relative, are on the brink of extinction, making it likely that we will soon be the only ones left of the ape family. Any species can potentially reach a time when it can not longer cope with the struggle against its environment. (Today, in zoos and other controlled environments, we maintain a few species that would be extinct without our intervention. But who would intervene for us if we needed such a lifeline?)

Extinction can be a slow and natural process, or it can be relatively sudden, as with climate change, meteorite impact, and disease epidemics. Nowadays, meteors are popular as a hypothetical cause of dinosaur extinction, but could a virus have wiped them out? We will probably never know. To answer this question, we would need to map viral genetic material after finding its ancient remains in a fossilized dinosaur. So far, we can't even isolate the genetic material of dinosaurs, much less their viruses.

In fact, most molecular biologists agree that RNA or DNA more than about 50,000 years old has disintegrated completely. Or it has become hopelessly contaminated by unrelated material—for example, DNA from skin cells sloughed by molecular biologists. One researcher, looking for ancient endoviral material in Egyptian monkey mummies, found DNA that was not from monkeys—but it was not viral DNA. It turned out to be from the chicken his Egyptian workmen had brought for lunch.

In fact, all attempts to obtain dinosaur DNA have ended in failure. However, in 1997, the research team of Svante Pääbo in Germany was the first to extract small pieces of DNA from the bone of a Neandertal man. Living in Europe from about 100,000 to 30,000 years ago, Neandertals were much like humans but not the same species. By comparing the Neandertal DNA with ours, Pääbo showed that we are not descended from the Neandertals and that, therefore, they are truly an extinct species. These results have been confirmed by a Russian-Scottish team headed by W. Goodwin.

What caused Neandertal extinction is a mystery. They may well have met their end through contact with humans, who left Africa about 40,000 years ago. These ancestors of ours first migrated to Israel, Turkey and Greece. Later, they may have lived in the Caucasus area for a long enough time to lose the melanin that protected their skin from the African sun. They eventually got to France, Italy, Spain and Portugal, where they co-existed with Neandertals for about 10,000 years.

The two groups must have interacted, but whether they had sexual relations is controversial. So far, the only evidence for this are the bones of an approximately 4-year-old child, found in Portugal, who died 24,000 years ago. Some scientists believe the child shows both Neandertal and human characteristics and thus indicates sexual relations. If so, that could suggest that Neandertals mixed enough with humans to be absorbed and not extinguished.

Most scientists doubt that interpretation, knowing that interspecies mating does not normally occur. But we are certain that Neandertals were pushed back to the south of Europe during the last 10,000 years of their existence. Scattered remains and other evidence show that the last Neandertal died in Portugal, not far in time or space from the 4-year-old child. Whether Neandertals were exterminated by humans, by a virus, or both, we may never know.

Certainly our ancestors were further developed and had better weapons than the Neandertals. If they transmitted a virus deadly to Neandertals, it would have been one that was not so deadly to humans. It could have been a relatively slow-killing virus, like the AIDS virus now decimating large parts of today's world. Or it could have been a flu virus, which can wipe out a population in a couple of months.

We might not believe that a virus could accomplish such a mass eradication, had we not seen the disastrous flu pandemic of 1918. Also, we know that a virus reduced the population of an entire continent in the 16th and 17th centuries. Most historians now agree that it was not the Spaniards alone who, during their quest for gold, defeated the New World people they erroneously called Indians. The smallpox virus the Spanish brought was the deciding factor. Writings of the time describe sieges in which many tens of thousands of Aztecs, Mayas and other indigenous Americans died mysteriously, even before the Spaniards began to attack.

There were apparently no outbreaks of smallpox in the pre-Spanish Americas, but the virus had been endemic in many parts of the Old World for thousands of years, killing at least a third of persons infected and sometimes a larger percent. During outbreaks or epidemics, the death toll from smallpox could reach 300,000 in a year. Since American Indians had never been exposed to smallpox, the death toll was probably more like one-half than one-third.

Now that smallpox has effectively disappeared, it is hard to believe how much people once feared it. "A pox on your house" was a serious curse, and "poxy" was a common insult, as in "a poxy knave." When smallpox erupted in their area, people fled if they could, as if a deadly flood were rolling toward them. Especially the rich fled, given their greater resources and mobility, but the disease struck all classes. Once infected, a victim's chance of survival was about 50/50, and survivors had ugly indelible facial scars. Their pitted faces could forever change their life. For example, a disfigured woman lost value in the marriage market and might never marry at all.

About two weeks after exposure to smallpox, an infected person would suffer a couple of days of headache, vomiting, and fever. The smallpox lesions would then erupt, always on the face. These sores remained visible until healed, and scars remained. The good news is that once the infection was over, it would not return. For people who had once had smallpox and survived, the chance of contracting the disease a second time was practically zero. The smallpox virus does not go into hiding inside the body like the AIDS or herpes viruses. Nor does the smallpox virus hide out in any non-human animals, though many animals have their own poxvirus. (In this book, *smallpox* refers only to the human virus and disease.)

The deadly smallpox virus was ultimately defeated by a competing virus, a much weaker smallpox virus, that gained the upper hand as a result of active human intervention. We conquered the killer virus by placing the only truly effective opponent in the arena: another virus. How did we do this?

Since one smallpox infection protects survivors from future smallpox, we realized that the trick is to cause that first infection with a relatively harmless virus. (Obviously it must not cause much harm, or replacing one smallpox virus with another would make very little sense.) The deadly smallpox cannot then gain access and cause its terrible disease. The weak virus would be removed from the body by the immune system, without leaving any visible sign of its presence except a little scar on the skin. Invisibly, for decades or longer, the immune system would be on guard for any renewed attack. It would be ready with specific antibodies or immune cells to any poxvirus that might come along.

Luckily for us, this trick could be accomplished because human smallpox exists in several strains, or variations. Some are deadly and some quite innocent. It was known for centuries that the serious illness, which killed more than 30 percent of its victims, was caused by *Variola major*. The mild form, *Variola minor* killed only one percent. It was also known that a case of the *V. minor* smallpox could protect against the *V. major* form.

At least by the 17th century and probably much earlier, the Chinese knew that deliberately infecting a person with the minor virus would cause a mild illness that protected against the major smallpox. So the crusts or fluid from *V. minor* blisters were eaten or applied to the skin. A popular method of what came to be called variolation (from *Variola*), may have been introduced to China from India as early as the 11th century. At first it consisted of using cotton wadding to absorb fluid from one person's smallpox blister, then placing the wadding in the nostril of the recipient. In the 17th century it became customary to cut a slight opening in the skin of the person receiving the fluid. A smallpox blister was pricked, a small cotton thread was saturated in its fluid, and this thread was placed in the cut. If all went well, the recipient would have a very mild case of smallpox and be protected for life from *V. major*.

The Chinese and others who practiced this art learned over the centuries which blister on which patient had to be pricked for the best variolation, or inoculation. They were aided by the very consistent external signs of smallpox lesions and scars. Even in very ancient times, everyone could easily see where and when smallpox had broken out. And that was not all. A smallpox expert could look at active lesions and tell whether the patient had a strong or weak chance of survival. It was strongest when the lesions were clearly separated from each other. When they blended together, on the other hand, the prognosis was much worse. And when smallpox patients had bloody blisters alternating with small red spots on the skin, they were doomed to die.

Obviously the blister fluid for variolation had to be taken from a patient whose smallpox lesions were clearly separated from each other. This would minimize the chance that variolation would cause more than a very mild infection. Variolation could thus be practiced without claiming too many accidental fatalities—if its practitioners had the expertise to distinguish between the various types of blisters.

By the early years of the 18th century, variolation had spread from the Far East to the Near East. It was widely performed in Turkey, where a prominent English noblewoman learned the technique in Constantinople (now Istanbul) from a group of Greek women who specialized in performing it. This woman, Lady Mary Wortley Montagu, was obsessed with smallpox because she had survived it and her brother had died from it. She promptly had her six-year-old son inoculated with a weak smallpox virus. Shortly afterward, on her return to England, she had her daughter publicly inoculated with the same virus. The procedure was rapidly adopted throughout England and, with some delay, in the other European countries as well.

Variolation was the first step on the way to eradication of smallpox, but many more steps lay ahead. Despite its success, opponents pointed again and again to the number of deaths caused by variolation. Being about one in fifty, this was definitely a problem, though quite an improvement compared to the one death per three cases that could be claimed by the aggressive smallpox virus.

Finally, the discovery of vaccination by Edward Jenner achieved the second big step. It arose from earlier observations that the painful little sores and blisters usually caused by variolation did not occur in people who were accidentally subjected to variolation a second time. Jenner noticed that a few people never had sores and blisters, even with the first variolation. It turned out that they all worked in the livestock industry. They were the farmers and milkmaids who had daily close contact with cows while milking them.

This gave Jenner the idea to use cow poxvirus instead of the weak human smallpox virus to protect against *V. major*. Commonly called cowpox, this virus has the Latin name *vaccinia* (from Latin *vaccus*, cow), so Jenner's procedure was eventually called vaccination. The word is now used for protection against all types of infections and illnesses that have nothing to do with cows. There is even talk of vaccination against Alzheimer's disease and certain types of cancer, though these last applications are still highly experimental.

Vaccination did not catch on immediately, however. Jenner had a hard time putting his ideas into practice because cowpox is quite rare. It caused perhaps 10 to 100 cases per year in England at that time, so he searched two years to find the right person with the right stage of cowpox infection: Sarah Nelmes, a farmer's daughter. On May 14, 1796, Jenner injected her blister fluid into 8-year-old James Phipps. Six weeks later, Jenner put the procedure to the test by injecting little James with smallpox. James got no sores at all. Nor did he get any when Jenner repeated the test a couple of months later.

The Phipps family must have been incredibly relieved. They took the chance, and we are lucky they did. Smallpox had been a killer for so long that its imagined death rate was no doubt even higher than its actual death rate.

It took a few decades, but gradually public opinion became convinced that vaccination was a good thing. The procedure offered as much protection as variolation but was much less risky, since only one person in a million would die from the resulting infection. A lingering problem was how to produce enough vaccine to vaccinate everyone who was at risk of contracting smallpox. Once this problem was finally solved in the 20th century, the World Health Organization (WHO) decided to attempt eradication of smallpox by means of large-scale vaccination. By that time, the late 1950s, smallpox vaccine was cultivated on a large scale, using cowhide.

Now, since the Somalian case in 1979, there have been no more cases of smallpox, and vaccination with cowpox has been discontinued. As a result, only people over 50 might have immunity against the disease. They can still show you a vaccination scar, but their immunity has probably faded or disappeared entirely, and among young people, there is no protection at all. This vulnerability is a problem because cowpox remains in nature, as do many poxviruses that could conceivably infect humans. And they are not the most pressing danger.

A few vials of smallpox virus are still preserved in laboratory freezers belonging to the U.S. and Russian governments. Some scientists argued against the dangers of such preservation, but others prevailed, thinking that the repositories of virus had great research value. To our knowledge, 400 strains are stored in the United States and another 200 in Russia.

It is also presumed that another substantial stock of the smallpox virus is hidden in North Korean missiles. Hidden stocks may exist in other countries too. These preserved strains could conceivably escape, causing an accidental epidemic, or they could be deliberately used for biological warfare. Such use of smallpox is not a new idea. In 1763, in what was then the British colony of America, Lord Jeffrey Amherst ordered that blankets dipped in smallpox blister fluid be distributed among hostile Indian tribes. The effects were devastating.

The fear of biological warfare using smallpox has risen sharply since September 11, 2001, when terrorists attacked the World Trade Center in New York City. The U.S. government and many European governments are quickly producing smallpox vaccine for their entire populations. They are even considering vaccination for the worldwide population. However, vaccination does not work against every virus.

For example, a group of Australian researchers discovered in 2000 that they could convert a totally innocent mouse poxvirus into one that kills mice in a couple of days. Worse still, little could be done to fight this very nasty creation. The majority of mice were still dying from the infection even after vaccination, which normally protects every human or animal against a poxvirus. A better weapon for biological warfare can hardly be imagined.

Besides the lethal smallpox virus, the vaccinia virus used for smallpox vaccination is also preserved in many countries. The belief that it is still obtained from an actual cow is based on a persistent myth. Genetic research shows that the vaccinia virus of cows and the variola virus of humans are closely related. Probably both are descendents of a rodent poxvirus, since rats and mice have often infested cowbarns and human shelters. In any case, the current vaccinia virus seems to be a combination of the original cowpox and a human virus that arose only a century ago—about the time vaccination started.

This is plausible because vaccinations must have been performed accidentally in at least a few people who were already infected with the human smallpox virus. Some of their cells could have been co-infected with both viruses. Once a virus enters a cell, later entry by another virus—superinfection—is almost never possible. But co-infection can occur when two viruses arrive at the entry molecule with just the right timing. Once they are both inside, they might mix genes, producing the hybrid or recombinant virus that by now has edged out the true cow poxvirus.

Another risk is that some form of the smallpox or vaccinia virus will manage to hide in a pet or an animal used for human consumption. This seems to have happened already in Brazil where in 1999 an epidemic occurred among cows and the farmers who were milking the cows. It was discovered to be caused not by a true cow poxvirus but, instead, the vaccinia virus with which the Brazilian population had been vaccinated more than 20 years before. In other words, it was a cowpox that the cows got from humans. There were a few infections with this virus during the WHO vaccination campaign, but now there seems to be a new reservoir of the vaccinia virus in parts of South America.

A similar thing seems to have happened in India, where an epidemic of smallpox erupted among people due to a virus that came from buffaloes. It was not a buffalo virus but the vaccinia virus that had been used for smallpox vaccination in India.

Aside from such hybrids, the animal poxviruses that exist in nature—the ones we know about—could not cause new epidemics in people except after recombination and/or long adaptation. This is possible but not likely. A good example is the myxomatosis virus. This rabbit poxvirus has been used to reduce the rabbit population in Australia, which is overrun by millions of rabbits that eat the crops. The origin of this plague is somewhat ironic. The rabbits are descended from rabbits once carried on many ships, to provide fresh meat on long voyages. They sometimes escaped to land or were deliberately left in remote locations, such as Australia, to provide emergency rations for stranded seamen. This idea seemed to make sense because rabbits multiply so quickly. In Australia, however, they had no natural predators and multiplied at an unprecedented rate.

Unlike the smallpox virus, the rabbit poxvirus is transmitted by the bite of mosquitoes and flies. Two weeks after a rabbit has been bitten and infected, it dies a horrible death from the disease, which causes swelling of its eyes and ears due to oozing lumps under the skin. The myxomatosis virus was discovered by Sanarelli in 1898. In 1918, the Brazilian physician Aragao suggested its use to the Argentine and Australian governments, both facing a plague of rabbits. The idea was slow to win supporters, but in 1951, it was finally being tested by researchers on an island off the coast of southeastern Australia.

Suddenly the virus escaped to the mainland by way of infected mosquitoes. In the Murray Valley, rabbits began dying in droves. A human outbreak was feared, especially when people in the same valley showed signs of a new illness. It too was transmitted by mosquitoes, and everyone thought that the rabbit poxvirus had struck humans. This turned out not to be the case. The human epidemic was soon shown to be caused by what is now called the Murray Valley encephalitis virus.

As so often happens, this scientific fact made little impression. Rumor and panic continued until the two most important Australian researchers of the myxomatosis virus, Burnett and Fenner, finally decided to inject themselves with the rabbit virus. Both remained healthy, thus proving that the myxomatosis virus could only infect rabbits.

Meanwhile, it went on killing more and more rabbits over a widening area of Australia. But after a few years, something very strange happened. A study conducted between 1956 and 1981 found that fewer and fewer rabbits were dying from the virus, even though the vast majority continued to be infected. While mortality had initially been 87 percent, it was 60 percent toward the end of the study period.

It seemed that the rabbits were building a resistance to the virus, which meant that the virus itself had to become more aggressive to survive in rabbits. Soon the rabbits were more often becoming ill but were dying less rapidly. The disease lasted longer, so there were more rabbits walking around with large quantities of virus in those lumps under the skin, so the virus could therefore be more easily picked up and spread by the mosquitoes to even more rabbits.

The longer duration of rabbit illness was of the utmost importance to the virus. After all, it somehow had to get through the winter, when cold weather kept the mosquitoes from coming around. For the virus to survive the winter, the rabbits had to survive the winter. The result was good for rabbits and virus, which both lived longer, but farmers got the short end of the stick because the rabbits kept eating their crops. (Whenever humans think they can evade the laws of evolution, the outcome has all too often turned out to be a let-down.)

Why do we care about myxomatosis? In nature, the evolution of any parasite tends toward a balance that lets the parasite thrive without causing too much misery for the host. A dead or dying host is usually not the optimum host. So a virus that is too virulent for its own good may evolve to be less virulent—though it rarely becomes entirely innocent.

Of course, like a virus headed for its favorite cell, a virus population doesn't do this deliberately. In a viral population, some individuals are just naturally more lethal than others; and if the too-lethal ones kill the host too fast, they don't produce as many of their kind; they gradually die out as the kinder viruses flourish. Such a change would take eons for humans, who take many years to reach reproductive age. It takes viruses a few weeks or months to evolve, since they reproduce early and often—several times a day. The myxomatosis story shows how a balance can evolve to allow *populations* of host and virus to survive even as *individuals* still die from the virus.

However, Australian farmers were unhappy with increased rabbit survival. They pressed for another solution, as did persons concerned that the rabbits were suffering greatly and dying a very cruel death with myxomatosis. Since 1997, a new virus for getting rid of the rabbits is being tested in Australia. Known among virologists as *calicivirus,* it was mentioned in the chapter about plant viruses. It is spread not only by insects but by direct contact and causes less suffering in the rabbits.

Ironically, rabbit-related research in Spain has recently yielded a weaker myxomatosis virus that protects rabbits from the lethal form. Spain has very few rabbits, and if their survival is threatened, the animals that prey on them could become extinct. This situation shows how local environment and living conditions—which differ so greatly between Spain and Australia—are crucial to the evolution and survival of each species and also to evolution and survival of all its predators and parasites. The Spanish would like to let their weaker myxomatosis virus escape into the wild, hoping it will edge out the stronger one—much as cowpox edged out smallpox.

Finally, we must wonder about the animal poxviruses we've never met that live in primeval forests and other unexplored areas. Particularly those that infect non-human primates could conceivably threaten human primates, as was shown in 1996, when people became ill with smallpox in a remote region of Zaire's jungles. Six of them died. The epidemic was caused by a previously unknown monkey poxvirus. In the Congo, an isolated case of a monkey poxvirus infecting human occurred in 1970, but 1996 saw the first outbreak in which the monkey infection was transmitted from one human to another. Monkeypox virus hits about one in ten family members of affected households. Clearly this attack rate was and is too low to sustain monkeypox in the human population. The monkeypox virus needs a more susceptible animal reservoir to survive other than primates, be it monkeys or man. The monkeypox outbreak in 12 villages in the Congo was most likely due to eating of raw infected meat from wild-caught animals. Animals shown to be infected during this African outbreak included Gambian rats in addition to squirrels and a domestic pig.

Until 2003 monkeypox infections were exclusively seen in Africa. In that year an epidemic of monkeypox occurred simultaneously in six states of the United States (Illinois, Indiana, Kansas, Missouri, Ohio and Wisconsin). Between May and July of 2003, a total of 71 cases occurred, all after having bought a prairie dog as pet. All animals originated from a single Illinois animal distributor. During the period the prairie dogs were raised in the Illinois facility a shipment arrived at the end of April of imported African rodents. These rodents, all from Ghana, were imported to the United States by a Texas animal distributor at the beginning of April. Gambian rats and squirrels from this shipment were shown to be monkeypox-positive. This very recent event shows that today nobody in any corner of the world is safe from almost any virus infection.

Viruses of rural Africa or Asia may show up in the United States or Europe because of the intense traffic of animals and humans at any moment in time. What once was a risk only for travelers to high-risk areas, where a virus is endemic and has its animal reservoir, is now a risk for almost anybody wherever he or she lives. All because of a diaspora of hosts, not of the virus itself.

So the smallpox virus and its close relatives are still among us in one form or another: in lab freezers, in the animals of rural areas, and in primeval forests. It is, in fact, very doubtful that this virus—or any human virus—can be truly eradicated without human beings also disappearing from the face of the earth. The WHO eradication project was successful and we are far better off today than before it occurred. But quite possibly we have not heard the last of smallpox—even if it is never used as a weapon—and will have to resort again to the protection of vaccination.

Interestingly, there are a few people walking the earth who may need no protection. When exposed to human smallpox, they incur little or no infection. As mentioned in chapter 8, very recent research has shown that some poxviruses use the same receptor, or cell entry molecule, that is used by the totally unrelated HIV. The molecule is called CCR5. Just a few years ago, AIDS research showed that mutation of the two genes coding for CCR5 keeps the AIDS virus out. And if only one of the two genes is altered, HIV can enter but causes a less serious infection. This may apply also to smallpox.

In fact, we suspect that these few people are protected from AIDS, a new disease, because their ancestors were survivors of smallpox, a very old disease. Those ancestors just happened to carry the CCR5 mutations, which have been passed down through generations in this small group. The mutations have been found only in Europe or in areas where there are descendants of European colonizers.

We must conclude, at this point, that eradication of a virus without eradication of its host may be only a dream. And the only thing more difficult than wiping out one virus is wiping out a whole group of related viruses, such as smallpox, vaccinia, and monkey poxvirus.

Eradication of a virus may remove it from the earth and, with it, the symptoms of illness it causes—at least in the short term. But we must wonder if vaccination should be halted, as was decided for smallpox. The continued use of a weak virus whose infection can provide resistance against a lethal virus might be a safer strategy, in the long run, than halting vaccination after presumed eradication. Simply to decrease the seriousness of an infection may be the more realistic strategy, although less spectacular.

Attempts at eradication—with its ultimate consequence of stopping vaccination—may bring us a false sense of security. It may allow a dangerous virus to resurface unexpectedly in a more sinister form than ever before. It may be because one host species eats the other or the host destined to be a pet is transported over long distances, as happened with monkeypox; because one host species purposely infects another host species, as has happened with rabbitpox; or because individuals of the host species fights others of the same species, as may happen if terrorists weaponize smallpox.

10

RAIDING THE WILD FOR DELICACIES

The SARS Virus

The SARS virus has its origin in Guangdong Province in China—the home of all those dangerous flu viruses in chapter 1. Every generation of virologists since the First World War has feared that a 1918–like flu virus would re-emerge one day. So far it has not, but when the first cases of SARS emerged, everybody suspected that finally a new strain of deadly flu had shown up. When influenza virus was excluded as the cause of flu-like SARS, the race was on to identify the true culprit of this fierce epidemic.

Several weeks before the race began, an old professor arrived in Hong Kong for a wedding. He came from Guangzhou, the capital of the neighboring province of Guangdong. When he checked in at the Metropole Hotel, on February 21, 2003, he took a room on the ninth floor. He had a fever and dry cough and soon had infected seven other people on the ninth floor, perhaps by sharing the elevator with them. Some of these people then flew from Hong Kong to Singapore, to Hanoi in Vietnam, and to Toronto in Canada.

Suddenly the experts who watch for disease outbreaks saw that a new global epidemic was born. By the end of May 2003, more than 8000 people had been stricken. The death rate, averaged across age groups, was about 10 percent. By June, the epidemic had slowed but was still spreading in mainland China and Taiwan and, to a lesser extent, in Hong Kong and Canada. On July 5, 2003, the epidemic was declared "over" by the World Health Organization. But by the start of 2004, SARS may have resurfaced in China.

At first the disease had no name, but soon it was called "severe acute respiratory syndrome," or SARS. Clearly it is a "severe respiratory syndrome," and to physicians, "acute" refers to the first, or primary, attack of an infectious agent. An acute disease is often brief and self-limited. It simply establishes the infection, which may or may not recur; if it does, the infection is "chronic" or "secondary." An acute disease is not always severe or even noticed, but in this case it certainly was.

SARS could be deadly because its initial fever and dry cough often led to congestion of the lungs that in its end stage made breathing impossible.

At first nobody knew where SARS came from, and the facts were slow to come out. Eventually they showed that before coming to the Metropole, the professor had been treating patients with a deadly pneumonia in Guangzhou. He had a mysterious pneumonia that had already killed many Chinese in Guangdong, once known as Canton. The disease had emerged about six months before, but the news had not spread outside the province. Apparently the first patient was a cook at a restaurant in Shenzhen, Guangdong, who routinely prepared a wide selection of animals caught in the wild. The cook was admitted to the Futian Hospital in Shenzhen on August 20, 2002. He had infected his wife and two of his sisters at home and would infect a number of doctors and nurses who took care of him in the hospital.

This pattern of SARS spread turned out to be universal, repeating itself in Hong Kong, Singapore, Hanoi and Toronto. Everything pointed to a virus that is transmitted through close person-to-person contact. Such contact is not as close as exchange of bodily fluids, as in sexual intercourse, but when a nurse takes care of a SARS patient, there is close contact. When a doctor does a physical examination, there is close contact. When people share the same living space, they have close contact. Kissing and hugging are close contact; sharing eating and drinking utensils is close contact; even talking to each other at a distance of less than three feet is close contact.

Such contact means not only touching the skin of infected people but also touching objects that they have touched, from door knobs to elevator buttons. These people and objects are contaminated with aerosol droplets containing the virus, so touching them is infectious if you do not quickly wash your hands before absent-mindedly bringing them to your nose, or eyes, or mouth. The best protection against SARS is to wash hands frequently with soap and water; to avoid touching your eyes, nose or mouth with unwashed hands; and whenever possible, to urge people around you to cover their coughs and sneezes with a tissue.

Close contact does not mean passing an infected person on the street or sitting across from them in a spacious waiting room. It does not generally include public encounters, unless people are crowded very close, as in an elevator. So masks are not normally needed by uninfected people going about their routine public business.

However, masks certainly help an infected person to avoid spreading the virus to others. And when hospital personal that care for SARS patients on a daily basis make a point of wearing masks, gloves, and gowns —and washing their hands frequently—they are protected completely against SARS.

People with SARS are most infectious during the time that they show symptoms: fever and a dry cough. There are apparently few asymptomatic cases. Most infected with SARS seems to have symptoms, which means that we can always tell when they are infectious to others. On average, they will not show symptoms for about six days after exposure. Then, after symptoms disappear, today's sensitive tests may detect virus in their throat or stool for up to two weeks, but the virus is no longer dangerous.

A surprising finding is that Asian people seem to be more readily infected by the SARS virus. This may suggest that Asians are particularly vulnerable to SARS, but the role of ethnicity is not yet clear. What is clear is that age matters, as it does in many diseases. SARS patients older than 60 years of age have about a 55 percent chance of dying. For young adults, the chance is 7 percent. Children from age one to sixteen can get SARS, but the course of infection appears to be mild for that age group.

Obviously, the elderly would benefit most from SARS protection. As soon as a vaccine or an antibody might be available, all persons 60 years and older in SARS-endemic areas or traveling to such areas would be prime candidates for vaccination or preventive antibody treatment. As already explained, vaccination involves injection of a virus that is weaker than the virus to be prevented; the vaccine virus cannot cause infection— or causes only very mild infection—but can keep the more aggressive virus away. Preventive antibody treatment involves the injection of antibodies to a certain virus; they supplement or stand in for a person's own antibodies to the virus. Their protection is not as long-term as vaccine protection, but they often work faster than a vaccine. Sometimes they are used in combination with a vaccine.

The three options are well illustrated by the Ebola and hepatitis A virus infections and rabies. The first two, like SARS, are spread through close contact. Ebola cannot yet be prevented with vaccine or antibodies. Again like SARS, its containment is completely dependent on quarantine (in addition to urging non-quarantined people to avoid risky situations). For hepatitis A, however, a highly effective vaccine is readily available in endemic areas. It is provided on a global level to travellers to such areas. In lieu of vaccination, antibodies can provide short-term protection if given a day or two before exposure to hepatitis A or within two weeks after exposure to the disease. As for rabies, as soon as a person is known to be infected, they are quickly given antibodies together with vaccine.

With SARS, as with Ebola, there is not only the problem of how to produce effective drugs for prevention or therapy. There is also the problem of who will do this. Most pharmaceutical companies are not motivated to invest in the necessary research and development unless a disease is much more widespread than these two. Drugs may eventually be forthcoming, however, and meanwhile we are fortunate that isolation and quarantine measures work so well. From all information available, they work very well if consistently enforced, particularly if applied early in an epidemic.

Quite possibly, if such measures had been taken when SARS first appeared in Guangzhou—or even when it first appeared in Hong Kong—the disease would never have left China. The good news is that, because it spread, we are now aware of a new virus that might have gone unnoticed. We have learned a lot about it, which will help us when it causes trouble in the future.

Every clinician and researcher confronted with SARS predicted a virus to be the cause. Was it an old one up to new tricks—or was it an entirely new one? Samples from SARS patients contained no virus (or bacteria) known to cause the SARS-like pneumonia. Thus a new agent was suspected, and researchers from the United States, Europe, and Asia rushed to see who could find it first.

The three competing groups very skillfully used the several methods available to find a new virus. It must first be grown on cell culture, because its preferred cells, nutrients, and growth patterns give clues to its identity. Then it must be isolated and purified: the new virus must be rigorously separated from anything else that may have grown in the culture. Only then can it be identified.

Growing viruses is not always easy. They need particular cells in which to multiply, and some are very fussy. Some human viruses only grow on human cells, but other human viruses will grow on a wide variety of cells, including those from birds, dogs, or monkeys. SARS virus did not grow on most cell types routinely used in clinical virology labs—until fetal rhesus kidney cells were used. Known as VERO, this culture began to grow the virus two to four days after it was touched with sputum or throat swabs. The cultures were showing clumps of rounded cells, indicating that a virus was present. The first sign that a virus is multiplying in a cell culture is often a change in cell morphology, or shape, followed by cell death.

The virus grew to high titers and was soon identified as a coronavirus, first by the Hong Kong University group in the Queen Mary Hospital. Virologists led by Malik Peiris isolated the SARS virus in March and published their observations on April 8, 2003, in the British journal, *Lancet*. Publication in such journals usually takes months but the process was vastly accelerated in response to the SARS emergency.

The Hong Kong researchers were quickly joined in their conclusion by the US group from the Centers for Disease Control, headed by Thomas Ksiasek, and the European Union group from several German, Dutch and French universities, headed by Christian Drosten. Both the U.S. and EU groups published their observations on April 19, 2003, in *The New England Journal of Medicine*. This journal, like *Lancet*, is in the top tier of prestigious and widely read medical journals. They are refereed journals, meaning that each potential article is referred to several experts in the authors' field, and final acceptance is based on their opinion.

The virus isolates from all three groups were identical. The Peiris group reported two isolates, the Ksiazek group identified one, and the Drosten group identified one isolate. All three groups identified a coronavirus as the putative cause of SARS. This was surprising as coronaviruses were not usually dangerous, nor did they grow readily on VERO.

Interestingly, Peiris was initially convinced that SARS was caused by a flu virus. He had come to Hong Kong in 1995, just two years before an avian flu outbreak killed six of the eighteen people it infected. It was an H5N1 virus from Guangdong, and the people were infected from birds, not person-to-person contact. However, when Peiris heard in January

2003 about the mysterious pneumonia in Guangdong, he feared that H5N1 had learned to spread among people. He knew H5N1 had reached Hong Kong, because in December of 2002, several geese in the parks of Hong Kong had died of H5N1 infection. So in February of 2003, the first SARS case understandably looked like it might be H5N1.

We are fortunate that what turned out to be the SARS epidemic did not coincide with any kind of flu epidemic. Since no flu was around in spring 2003, "true" SARS cases could be more easily spotted. Also, the healthcare system was not as overloaded as when two epidemics coincide, especially when one benefits from quarantine and the other does not.

The system was overloaded enough with SARS alone, especially since the disease was of particular danger to healthcare workers. Even in Toronto, a large and modern Western city, physicians and other workers were so demoralized that the Canadian government offered US physicians remuneration to assist. Several hundred physicians did so, but some felt they had to conceal their mission to avoid problems. Even in such a modern city, many people had unreasoning fears of the disease.

While hospitals coped with SARS itself, researchers probed the coronavirus to prove definitively that it caused SARS. Just because someone is sick and a pathogen is found in her body, the pathogen cannot be assumed to be the cause. In the same way, if a robbery is committed and someone is found near the scene, that person cannot be assumed to be the culprit.

More than 100 years ago, the German bacteriologist Robert Koch devised "four postulates" that are still used to prove causality. As mentioned in chapter 3 on prions, the suspected agent must first be found in every case of the disease. Second, it must be grown in culture and purified. Third, inoculation of the pure culture into experimental animals must produce the same disease. Fourth, the suspected agent must be seen in, and recovered from, each sick animal.

In a similar manner, SARS researchers looked for antibodies, which were far from discovery in Koch's day. Specifically, they looked to see whether antibodies to the SARS virus were absent before people had SARS and present when they had recovered from SARS. The easiest way to do this was to attach SARS-infected VERO cells to slides (using uninfected VERO cells as controls) and then to see what antibodies have appeared in the course of the infection. These are revealed by a coloring technique called immunofluorescence, by which SARS-specific antibodies are highlighted.

All three groups showed that SARS patients developed SARS-virus specific antibodies. Moreover, they found these antibodies in every single one of the SARS patients tested. The Peiris group additionally showed that all patients without SARS but suffering from other respiratory infections lacked the SARS-virus antibodies, as did healthy blood donors.

These findings clearly indicated that the SARS virus was new, that the virus was strongly associated with the specific symptoms of SARS, and that the relationship of the virus with SARS was more than coincidental.

The next step was precise identification of the virus. Was it a totally new coronavirus or one that had suddenly changed? Ordinary coronaviruses cause about 10–20 percent of human colds and sometimes diarrhea, but nothing more serious. Results obtained by Peiris strongly indicated that the SARS virus was not a "changed" human virus. Could it be the changed virus of a bird, a mouse, a cow or a cat? Chickens get bronchitis from a coronavirus; cows, pigs and dogs get diarrhea.

The SARS virus had to be confirmed as a coronavirus and then placed somewhere on the coronavirus family tree. The simplest way to do this was to inject the cell cultures with antisera specific to the various groups and subgroups of coronavirus. Each antiserum contains just one type of antibody, and when various antisera are introduced, in turn, to samples of the same cell culture, the results can be compared. Whichever antiserum colors the cells identifies the virus infecting those cells, right down to their twig on the family tree.

SARS researchers tested the cultures with antibodies against every respiratory virus they could think of, but none colored the cells. So they took the next step: electronmicroscopy. Virus particles are too small for a normal microscope, but the electron microscope showed that the size and shape of the mystery virus definitely made it a coronavirus. The same virus was not only seen in slides made with samples of cell culture but also slides made with sputum or throat swabs taken from SARS patients. Not only that: only coronaviruses were consistently recovered from SARS patients—more confirmation that this particular coronavirus was the cause of SARS.

Some readers may be wondering whether the researchers would ever be convinced. In science, a finding is not accepted as fact until every possible test has been used to disprove it. The process can take years, and sometimes a fact is finally "proven," only to be overturned later. Sometimes a fact turns out to be unprovable, in which case it is sometimes accepted anyway, as with the law of gravity. Journalists and the public can find scientists hard to pin down, but good scientists are just being very careful.

It can take years to show unequivocally that a certain agent is the cause of a certain disease. Koch's postulates have served very well, but they were designed to identify bacteria as the cause of disease in a time when both viruses and antibodies were unknown; they are still the cornerstone of such research, but refinements have been added over the years.

In the 1930s, Rivers modified them to prove a virus to be the cause of infection. He postulated six requirements. The first three were quickly met by all three groups of SARS researchers: isolation of the virus from diseased hosts; cultivation of the virus in host cells, and use of filters to prove that the agent is smaller than bacteria. These findings were supported by observing virus growth in monkey cells and virus size and shape on electron microscopy.

Three Rivers criteria remained: production of comparable disease in the original host species or a related species; re-isolation of the virus from that experimentally infected host, and detection of a virus-specific immune response following the experimental infection. This last response should be comparable to the immune response in natural infection.

The credit for meeting Rivers' last three postulates for SARS goes to the group of Albert Osterhaus from the Netherlands. The Dutch team administered a few drops of culture fluid containing SARS virus obtained from VERO cell cultures into the eye, the nose, and the trachea of two rhesus macaques. Both animals developed SARS within a week, just as humans do. Both animals excreted SARS virus from throat and nose, beginning on day 2. One monkey excreted up to day 4 after experimental infection, the other until day 10. Both monkeys developed antibodies to the virus after two weeks. When the monkeys were euthanized after 16 days, autopsy revealed lung pathology exactly as was seen in the lungs of SARS patients that had died from the infection.

Finally, the Dutch group found that once the monkeys were SARS-infected, superinfection with agents like human pneumovirus and *Chlamydia pneumoniae* did not alter or worsen the course of infection definitively showing that the "SARS coronavirus" was indeed the sole cause of SARS.

All things considered, we have been quite lucky with this new disease. Less than two months after the outbreak was known to the world, the causative agent was definitively identified and diagnostic tests were developed. Now, every person with fever and a dry cough can be tested for the SARS virus itself and also for antibodies to the SARS virus. This means we can quickly identify people with SARS—and make sure they do not have flu—even in locales remote from "hotspots" like Toronto, Taipei, Hong Kong, or Beijing, and take the necessary measures to prevent the SARS virus from spreading.

We were lucky also in that the SARS outbreak was easy to see, because its victims always show symptoms. Recognizing an outbreak is often not so easy. An outbreak is a sudden increase in the rate of transmission of an infectious agent. In the case of re-emerging infections, the public health system is already aware of its symptoms. With an emerging infection, awareness must come over time. If symptoms are subtle or long delayed in appearance, recognition can take years, as it did with AIDS.

Our success with SARS shows that in the 20 years since AIDS arose, virologists have enormously improved their ability to identify and characterize a virus. The process can now be done in weeks, even with the setbacks that are inevitable in any venture that forces people of many cultures to work together. Admittedly, the SARS virus was not subtle and did not hide itself too well. But the recognition of this completely unexpected epidemic is still a great victory for the global public health system.

The English virologist Philip Mortimer recently postulated, in analogy to Koch and Rivers, five requirements for detecting and resolving outbreaks of infectious diseases. They apply whether the outbreak is an emerging or re-emerging disease: (1) there is a discernable pattern of time, place, and action. As Mortimer puts it, "An outbreak, like a story, should have a coherent plot." (2) there is a consistent mode of spread, usually from a shared source of infection. (3) those affected are among those exposed to this source, and not among any persons not exposed. (4) there is evidence of a unique microbial strain. (5) when the source of the implicated micro-organism is contained or eliminated, the outbreak wanes.

These requirements may seem obvious, but in the middle of a storm, it takes a prepared mind to grasp the situation. The fourth postulate demands the most of virologists, who must discover whether the same virus is causing all cases in an outbreak. Here they are assisted by an army of molecular virologists, who bring weapons borrowed from the genome scientists. The key is to decipher the complete genetic code of the virus.

With SARS, this feat was accomplished in a month or two. They had help from the genome centers of the world, which can rapidly determine the genetic code of DNA samples from tissues or cells of living creatures large and small, and also the code of bacteria and viruses. Each of the three competing research groups used the technique of polymerase chain reaction to obtain short stretches of coronavirus RNA. The stretches were long enough to enable comparison of the SARS virus to other members of the coronavirus family. The virus did not look like other human coronaviruses; it looked closest to mouse or cow coronaviruses, but not so close that either was thought to cause the SARS epidemic. The final judgement awaited revelation of its complete genetic code of about 30,000 base pairs.

On May 1, 2003, a Canadian group and a US/EU partnership separately reported in the journal *Science* the complete genome sequence of the SARS virus. The two isolates they had studied had virtually identical genomes, showing that a single virus strain caused the SARS outbreak. Not only that, but the virus was shown to be unique among coronaviruses. It was nothing like any of the known coronaviruses.

On May 9, 2003, the Genome Institute of Singapore published the sequence of 14 SARS coronavirus isolates from the same local epidemic. They showed the SARS virus to be new, because the isolates were so much alike. (If it were not new, strains would have had time to become more heterogeneous by means of the inevitable mutations.) Apparently the 2002–2003 SARS epidemic was truly the first appearance of this virus in humans. It had not been circulating for years among people in some remote part of China, where it might simply have gone undetected.

The next question was: Is this new and unique virus a recombinant virus, one-half chicken virus and one-half cow or mouse virus, for example? Most scientists believed not. They assumed that this unique virus originated from a not-yet-identified animal reservoir, most likely an animal living in the wild, because every SARS-infected person has a serious respiratory disease, whereas no wild animals were known to have such disease. Usually when a virus jumps, it comes from a host in which it is comfortable and causes no disease. It then finds itself in a host where it is uncomfortable and must work hard to survive, reproducing to disease-producing levels. Thus a jumping virus usually causes disease.

In contrast, even a newly minted recombinant virus might not be quite so uncomfortable in humans. Part of it would probably have come from a human virus. If it caused trouble, it would not cause as much trouble as we see with the SARS virus.

If the SARS virus had recently jumped from an animal to humans, what animal was it? The first SARS cases in Guangdong were food handlers and cooks, which points strongly to an animal that is eaten by the

Chinese. Chickens and pigs were the logical candidates, but both turned out to be negative for the SARS virus. So the hunt for the SARS virus reservoir moved to the wild-animal market, which is very active in Southern China.

The Chinese enjoy many delicacies from the wild and are willing to pay large sums for them. Once researchers began to examine such animals, they quickly found a coronavirus extremely close to the SARS virus in masked palm civets *(Paguma larvata)* and a raccoon dog *(Nyctereutes procyonoides)*. The viruses in these animals, both sold at a market in Shenzhen, were closer to the SARS virus than any other animal or human coronavirus, clearly belonging to the same unique coronavirus family.

A genetic match of viruses from such genetically divergent species as humans and civets strongly suggests a zoonotic transmission event: an animal disease that is passed to humans. Of course, the virus could have jumped from humans to civets, instead of the other way around, but the jump is most likely from animal to human since civets are not harmed by the virus while humans are. Moreover, no human has been found to carry the SARS virus or SARS virus antibodies except those who have had SARS infections (or their close contacts). If it were a human virus, it would most likely be found also in a number of symptom-free individuals.

More than 10 percent of civet handlers at markets in Guangdong showed SARS virus antibodies, indicating they had recovered from a SARS virus infection. This percentage is much higher than anywhere else. As for the civets, at least half appear to carry the SARS virus. None of the animals from which it was isolated showed any disease, so the virus has a safe haven in these animals.

Civets are caught in the wild in Southern China, and transported to markets all over that area. This trade is a fertile breeding ground for new viruses and cross-species transmissions. If the SARS epidemic is definitely shown to originate from a cross-species transmission at the wild-animal market, the history of the AIDS virus comes to mind. AIDS originated from monkeys and apes that were eaten by humans in West and Central Africa.

The remaining question is: Why did the SARS epidemic emerge in 2002 and not earlier? The most logical explanations include ever increasing exposure plus a chance genetic event. Human exposure to civets has risen with the rise of human population and demand for these animals. A civet does not travel farther than the local market, but people increasingly travel all over. So an elderly professor, who decades ago might never have left his hometown, gets infected and goes to a wedding in Hong Kong, where he infects people who rapidly spread the virus all over the globe.

The presence of anti-SARS antibodies in the blood of civet handlers shows that infections had previously occurred; probably they were relatively mild and hardly noticed. No epidemic arose. But the virus must have changed. We know this because the SARS virus isolated from people in 2003 is somewhat smaller than its civet counterpart. It appears to have lost the genetic code for 29 amino acids. This deletion creates a revised virus structure that gives the human virus an advantage in its new "human" environment, apparently a result of the adaptation to humans. The gene might thus enhance the ability of the virus to cause harm, to spread faster, to infect different cells of the body, or to evade the immune response. The new human SARS virus might have a competitive advantage over the old "true" civet virus. For example, it can transmit from human to human, and the old civet virus could transmit only from animal to human.

There is one human SARS virus isolate that lacks the deletion and therefore lacks the human structure. Nobody knows which patient this virus came from, which is unfortunate. It could represent a direct civet-to-human transmission of the ancestral SARS virus.

The Chinese government is proposing to forbid trading wild animals for human consumption and to forbid restaurants to serve endangered and wild animals. After a new case of SARS was identified in early 2004, the Chinese government ordered the slaughter of all civet cats. This might sound like a heroic measure, but there are several reasons why it might have little impact.

First, the new SARS virus is already among humans, and it might have more reservoirs than just wild animals that people eat. Second, eating wild-caught exotic animals is a 5,000-year-old tradition in Southern China unlikely to be totally stamped out. Most likely, its prohibition will make the eating of civets even more of an obsession. Eating wild-caught animals will become an underground activity outside the surveillance of the health authorities.

For the Chinese, such delicacies are not only a status food but also medicine. Many rural people believe that eating masked palm civets reduces their chance of winter infections like colds and flu. Ironically this myth promises to increase, rather than decrease, such infections. SARS infections will now join all the colds caused by rhinoviruses and adenoviruses, not to mention the flu caused by influenza viruses.

Like these infections, SARS may be a largely seasonal problem. As noted by Don Burke, epidemiologist and virologist of Johns Hopkins University, human infections that are transmitted by coughing and sneezing tend to be strongly seasonal. During the winter they flare up; during the summer they wane. SARS may well follow this pattern, in which case, this new disease will not plague us all the time, and we will know when to watch for it.

Seasonality has not been fully explained but is partly influenced by changes in temperature and humidity and by the crowded indoor conditions that accompany cold weather. With SARS, there is the additional factor that civet is eaten to protect against winter. Burke predicts that each year, SARS will decline in summer but return with renewed force each winter.

Outbreaks originating from cross-species jumps of virus are rare and occur by chance, but chances increase when potential human hosts are repeatedly exposed to animals carrying viruses that could launch a zoonotic epidemic. Wild roaming animals are of little danger because they do not have repeated contact with humans. Our danger comes mainly from wild-caught animals. Often such animals are handled and eaten only by select populations, who are often too isolated to cause an epidemic.

Even wild-caught animals are a limited threat, and for better and for worse, the threat is shrinking. One day, all the wild-caught animals will have been caught and eaten, and all the wild roaming animals will be confined to zoos. Humans will then be threatened mostly, if not exclusively by the viruses of domestic animals.

SARS joins the ever-longer list of new viruses or viruses expanding their territory, like the AIDS virus, the West Nile virus and all others that threaten the health of more and more humans. Some come and go with the change of seasons, others take longer to re-emerge, like pandemic flu, and some become a permanent part of human suffering, like the AIDS virus.

EPILOGUE

Viruses are a threat to us more than ever before. We can tell from the viral diseases that are spreading like wildfire. And we notice it most when these diseases end in death, when they strike many people or when they arrive from somewhere unexpected. Viruses are invisible but they turn out to be everywhere. They are in the air, in the water, in the plants and animals which we eat. And it seems that these viruses have been seeking out people more and more often during the last few hundred years. The Dutch Nobel prize winner Paul Crutzen has labeled this period the "Anthropocene."

According to Crutzen, the Anthropocene began around the end of the eighteenth century when the concentration of carbon dioxide and methane on earth started to increase. It is the period of unrestrained population growth and the industrial revolution, and it is the period of the intensification of agriculture and cattle breeding. Almost half of the earth's surface is by now taken up by man and this share will increase significantly in the future. More than half of all potable water is being used by man. It is likely that all the various human activities will cause the temperature of the earth to rise by some 3 degrees Celsius this century, more than the increase due to the El Niño effect. The primeval forests are disappearing rapidly, which causes ever more species of animals to become extinct. In about fifty years, our nearest neighbors, the apes, will live only in captivity under the guardianship of the only primate still "living in the wild," man.

Where else other than among humans can a virus family settle down in the coming centuries? It seems that the migration from animals in particular to people as hosts has started, never to end again. One after the other, virus families are establishing themselves in our population, and we are trying to keep them out with all our might.

The first example was smallpox, one of the first viruses to leave deep marks on the history of mankind. Through vaccination, the virus was exterminated among people, but it lies stored in its original form in the freezers of superpowers such as America and Russia. In addition, smallpox—which in the past never had an animal pool—has now gone into hiding in a mutated form among cows and buffaloes. The AIDS virus has surfaced out of the dying world of apes, and it does not seem to be ready to leave man anytime soon. We can only hope that the AIDS virus will end up like the simple retroviruses, which ultimately became absorbed into their host's DNA, where they perform odd jobs for the host cell. But it has never happened yet that a virus of the AIDS virus family landed permanently in the host DNA, whereas the remainders of simple retroviruses can be found in our DNA by the thousands. The close symbiosis of the retrovirus with its host cell does exact a price, sometimes the cell degenerates into a cancer cell. In most cases, however, the retrovirus helps the cell to adapt to the environment. And sometimes the retrovirus provides assistance to other passing viruses in their struggle for survival.

Herpes viruses, however, go into hiding in specific cells of their host and they hibernate there all their lives. The human herpes viruses group does not have a pool in the animal world. They hide all their lives in the brain's offshoots and they only leave their place of hibernation in cases of stress. They spread inconspicuously, so that the host experiences as little distress from the infection as possible, to keep the host species alive. Herpes viruses infect man from cradle to grave. Most infections strike people while they are young, especially where conditions are primitive.

The development from poverty to riches that is being experienced by man, including improved sanitary and living conditions, threatens the existence of some viruses. Due to increased prosperity, the spread of viruses is occurring while their hosts are older, which appears to be more difficult and causes more illness. The solutions for this stalemate which viruses are seeking and finding are often surprising. The Kaposi's sarcoma (KS) virus spares women, which means that there is no danger to procreation, but it does cause cancer in men. Male HIV patients easily pick up the KS virus through oral sex. And when their resistance decreases, they contract KS—a proliferation of infected blood vessels.

Viruses guarantee their future by establishing themselves permanently in their host, as in the case of herpes viruses and retroviruses. Other viruses circle permanently around us. They do this by populating our lifelines, such as air, water and food. Airway viruses, such as the flu virus, literally circle around people. They are avian viruses that lead to sniffling and sneezing, which in turn spreads them around even more. Illnesses caused by viruses are generally associated with the manner in which viruses enter and then leave our bodies. The symptoms, whether coughing or diarrhea, contribute to spreading the virus. The immune disorder associated with an HIV infection increases the quantity of virus in blood and semen, thus making it easier for the virus to spread.

Influenza viruses teach us yet another lesson. These harmless viruses of wild geese and ducks evolve slowly in the cells of their natural hosts. This is fine in times of quiet and stability in the environment, but it is a disadvantage when they need to adapt to new circumstances. In animals such as chickens and pigs, and in man, the flu virus can adapt very quickly to changing circumstances. It is possible that the flu is transmitted to chickens and pigs, which provide people with meat, when pressure is exerted on the virus. The influenza virus then moves from chickens and pigs to man. As a result of the changing circumstances in intensive cattle breeding, the flu virus might very well in the future become more and more a virus of farm animals and people, thus becoming less and less a virus of birds.

Without food and water, people will die. So what would be more obvious for a virus than to hide in food and water, and then make its way into people like a Trojan horse? We live on fruit, vegetables, and meat. And also the animals which we eat live on the crops which we cultivate. There are plant viruses which cause harvests to fail and there are viruses which contaminate food and infect our intestines. Harvests fail due to infections with viroids. These are small infectious particles that cause disease in crops but infect wild plants without causing symptoms of illness. As is the case for people, population density seems to play a part again, this time for a plant population. The greater the population density, the faster the infection spreads and the more frequently illness strikes. The aggression of these viroids seems to be linked to their spreading, and this seems to be a general rule. Viroids usually only occur among plants. There is just one viroid-like particle in man which can only multiply with the aid of a helper virus and a protein stolen from the host cell. Viroids seem to have protein characteristics which lead us to surmise that these small particles had their origins in the RNA world, which is assumed by many to have preceded the DNA world.

There is a permanent controversy about the directions viruses are headed, which regards the question whether viruses originated from living cells or living cells originated from viruses. Both theories are based on a nucleus of truth. There once was an RNA world. In this world, replicating viroid-like particles were formed which led a feeble existence. Bacteriophages caused bacteria to fuse, and some of those bacteria enveloped viroids. These cells turned out to be viable, and they evolved first into single-celled and subsequently into multicellular organisms. Retroviruses introduced DNA into the RNA world and then contributed to the exchange of information between the RNA and the DNA worlds.

What could be a more beautiful supposition than that viruses no longer saw a future for themselves as independently living organisms and created their own host in order to be able to extend their lives to the end of time? The price was high of course, their independence was over. From that moment on, viruses would be totally dependent on living cells for their continued existence. This dependence manifests itself on many levels. Viruses can still maintain themselves in the environment, in the ground and in the water, but they can only reproduce in living cells.

Viruses spread together with the cell and also as virus particles without the cell as casing. The cell is essential for viral reproduction, but it is not at all or only to a lesser extent essential for its spread. And both viral reproduction and subsequent spreading are inextricably bound up with the survival of viruses. Since viruses can survive for a very long time in the environment as a viral particle, especially when it is provided with a protein mantle, a virus can do without a cell for a very long time. Some viruses, such as herpes viruses and retroviruses, go into hiding in the host itself, inside the cell. Other viruses find a hiding place in other hosts, such as domesticated animals. These viral groups can keep themselves alive in relatively small population groups, without needing to reproduce too quickly. These are the so-called latent and persisting viruses. In addition, there are viruses that reproduce like mad and fly or ride from one host to another in no time at all. In order for a virus to survive in a population, as opposed to surviving in an individual host, the particular virus must spread and reproduce rapidly. Something like jumping from ice floe to ice floe is required. In other words, the greatest contribution a host can make to the survival of virus families is to promote the rapid spread and reproduction of the virus. Rapid reproduction leads to high viral density in a host and this leads to large quantities of virus, which the host excretes by breathing out, by defecating and by engaging in sexual activities. As a result, large quantities of virus end up in the air, water and food, and on the host's hands and in his mouth—the host is plunged into a viral bath.

But the host must also be able to become infected with the particular virus and to reproduce it and then excrete it again. There are two kinds of barriers which a virus must surmount. The first is the species barrier, since most viruses are species-specific, and the second is resistance. The ease with which a person can be infected depends on many factors. One of these is the accessibility of the cell which has a receptor for the virus, in other words, where the virus can penetrate. It is much more difficult for an AIDS virus to reach a blood cell, on which this virus depends for its reproduction, than it is for an influenza virus to reach a pulmonary cell or for an entero virus or a polio virus to reach an intestinal cell.

The conditions under which a host lives, or is forced to live, contribute to the ease with which viruses are able to spread. It is generally estimated that more than two-thirds of all virus infections can maintain themselves best in backward communities of people who live in poverty. In this sense, poverty, undernourishment and unsanitary conditions help virus families to survive. These infections share two characteristics: people are infected while they are young and the few cases of serious illness caused by the virus occur only in a small minority of those who become infected. Good examples are provided by entero viruses and the polio virus.

Sewers are full of entero viruses and polio viruses that have been excreted with feces. By way of mouth, they re-enter people who live in circumstances where there is no distinction between waste water and water for drinking or washing. This is the case especially in developing countries, where sewers discharge directly into rivers from which drinking water is also obtained and in which people wash themselves and their clothing. In this situation, undernourishment has been shown to be an additional factor promoting the spread of these viruses. It became apparent in China that entero viruses thrive better in populations that have a selenium deficiency. It was found that people who had not yet contracted a virus infection were more susceptible to infection due to their selenium deficiency than people with a normal level of selenium. And infected people who had a selenium deficiency were found to excrete more virus, therefore being more contagious, than infected people who did not have such a deficiency. It then became apparent that the unpleasant side effect for the infected people with a selenium deficiency was that the larger quantity of virus particles in their bodies led to heart abnormalities. The viral aggression seen here, in terms of its harmful consequences to the host, stems from its survival instinct.

Viral aggression is a result of the resistance experienced by a virus as it penetrates the host's body or its cells. A virus can only allow itself to be harmless when it has no trouble spreading and the host is extremely susceptible to infection; in other words, when few viruses are needed in order for it to spread. However, when viruses experience resistance, only those viruses will survive which are able to overcome that resistance, and it appears that the simplest strategy to accomplish this is to increase the production of viruses. If a few thousand virus particles are unable to penetrate the cell or body, maybe a few million will be able to reach that goal. Brute force is apparently a popular option in nature.

People are not bothered by viruses but rather by the damage that viruses inflict. A good example of this is polio. The polio virus causes infantile paralysis. In developing countries, there were rarely if ever any cases of infantile paralysis and it was even thought that polio did not occur in poor countries. In reality, there is more polio virus in developing countries than anywhere else. But the infection takes place when those who contract it are very young, and it almost always runs its course without any signs of illness. In addition, such a person will then be protected against polio virus infections that may occur when he or she is older.

Improvements in living conditions reduce the risk that polio virus infections will occur in toddlers and pre-school children, but they increase the risk of an infection at a later age—in young people in their teens and twenties—and thus the risk of paralysis due to the polio virus. In fact, the last cases of polio in the western world took place among the upper middle class. We are now seeing this phenomenon repeated in the developing countries, where epidemics of polio virus infections with a serious outcome increased as living conditions improved. All-out efforts are now being made to rid the world of polio by means of a worldwide vaccination campaign. Sabin was the person who discovered that a weakened live polio virus, once it infects a cell, will protect that cell against attack by an infantile paralysis virus. This happens even in a test tube, without the immune system being involved. It gave Sabin the idea to flood the world with a live weakened polio virus, thus creating a barrier against the virus which causes polio. This is what is presently happening. Entirely as expected, we are now seeing the first cases of polio as the result of a mutated Sabin polio vaccine strain. The weakened polio virus, which is completely harmless, has difficulty spreading in nature and that is why virus strains get selected which are able to move somewhat more easily within a group of people. The mutations that enable the virus to spread more efficiently are also the mutations that enable the virus to penetrate the peripheral nervous system, thus causing polio.

Only viruses of domestic animals and viruses of people seem to be ensured of a future, and of course we must not forget the viruses of agricultural crops. There will not be many wild animals around in the future, or more precisely put, wild animals that really live in the wild. The small groups of wild animals that will remain in wildlife parks and zoos may very well be too small to offer a prospect for survival to most virus families.

One of the most deadly viruses that ever existed is the rinderpest virus, which is a virus of cows, sheep, and goats. The rinderpest virus is in the process of disappearing and, some 5000 years ago, it was transmitted to man. When it strikes man, it is called measles. Measles needs a group with a size of at least one quarter to half a million people to maintain itself. And this was the size of the population at the time of the Sumerian civilization around 3000 B.C., by which time all the large domesticated animals, specifically cows and sheep, had been tamed and were being bred for consumption. Large groups of hosts of the measles virus group came into proximity with each other there for the first time. And, in the case of man, it happened in sufficient numbers to allow the rinderpest virus to survive exclusively in man as the measles virus. Ever since then, measles has been entirely dependent on man for its continued existence since it is unable to infect any animals. It spreads through the air and it claims many victims particularly in developing countries. Undernourishment seems to play a role in measles as well; in fact, vitamin A deficiency turns measles into a deadly disease. In addition to a large-scale vaccination program against measles, recombinant rice containing pro-vitamin A could offer a solution here.

Viruses feel more at home in water than anywhere else. One liter of water contains 10 billion virus particles. These are bacterium viruses with only one single enemy, sunlight. When exposed to sunlight, bacterial viruses become less infectious and consequently reproduce less effectively. These viruses feel best in dark, deep water. There are more viruses in the oceans in summer than in winter. It seems that these viruses regulate the diversity of the bacterial population. When a bacterium family becomes too dominant, bacteriophages set out to restore equilibrium. Phages create living space for bacteria in distress. And again there is aggression, this time on the part of the bacterium which secretes toxin. Toxin causes one to contract illness from a bacterium. But is that the only purpose of this toxin? And why do phages carry that toxin from bacterium to bacterium? It is possible that host specificity plays a role here. A cholera bacterium without toxin can perhaps maintain itself in oysters but not in plankton, while a cholera bacterium with toxin can maintain itself better in plankton. And survival might very well be simpler for the cholera bacterium as a parasite of plankton than as a parasite of oysters. Phages which do not kill bacteria help the bacteria to survive, and thus they help themselves. But there are also phages which kill bacteria. These are often so bacterium-specific that they help one bacterium to survive by killing another. They guard, so it seems, the diversity in the population of bacteria.

Viruses of mammals might be able to help regulate the diversity in a population of mammals. It is easy to predict that this will become difficult if the mammalian population is restricted to domestic animals, people, and pests such as rats, mice and rabbits. All viruses of wild animals will then have to find a safe haven in one of these hosts, lion viruses in cats, wolf viruses in dogs, and simian viruses in people, but also mouse viruses in people and even bird viruses in people, and so on. There is not much choice.

If this transfer to another, more-or-less-related host goes smoothly, it will hardly be noticed. But if it requires a lot of effort, the new host may be confronted with a lot of misery. As an example, the AIDS virus cannot be transmitted easily from ape to man or from one human being to another, because we do not have white blood cells on our skin or on our mucous membranes. Lots of virus is needed to penetrate the white blood cells through small wounds or through the intact mucous membranes. In order to concentrate a lot of virus in the semen, virus production needs to be so high that it causes the immune cells in which the AIDS virus multiplies to die. And the immune disorder which is caused by the AIDS virus as a side effect of maintaining the quantity of virus will lead to the illness called AIDS, from which the patient always dies eventually if no medicines are available. It seems that viruses are continuously involved in clearing away hindrances that keep their fellow-travelers from spreading easily. They have an enormous arsenal of possibilities at their disposal and every virus family has found its own way of getting around obstacles.

Viruses make use of all kinds of escape routes, and they will continue to do so all their lives to evade the host's immune surveillance. Sometimes they quickly get away as soon as a new host has been infected; or they go into hiding inside the body; they dupe immune cells; they try to use genes that are look-alikes of host genes; they try to change their shape like chameleons; or they try to bribe their host.

This last strategy, no matter how reprehensible in other aspects, might well point the way to the future. For a virus, it may be best to persuade people to promote the spread of viruses rather than trying to counter it. For man as well, this might perhaps be the best strategy for living in peace, that is, without harmful consequences, with the many viruses which will be heading our way over the coming years from the animal world under threat. Vaccination with live viruses that not only supplant the harmful virus but also bring man something useful, for example, resistance against all kinds of ailments or other intruders. Vaccination combined with gene therapy.

Viruses destroy life and protect it at the same time. They are tight-rope walkers and equilibrium-keepers at the same time. In the Anthropocene era, nature is being reduced to a bacterial world, a world of all kinds of forms of primitive life, and the world of man with his domesticated animals and his pests. For viruses, there will be less to do in the future than formerly because there will be ever less biodiversity to be guarded. Although this may apply to the vertebrates, and the mammals in particular, it does not apply to the lower life forms or the bacteria. They seem to have been blessed with eternal life, and thus their viruses with them. The same might very well apply to the viruses of (vertebrate) animals, although they will have to make do with ever fewer host species.

This may very well result in ever more virus families becoming dependent for their survival on ever fewer host species. This in turn results in man and his domesticated animals, such as cows and sheep, as well as his pests, such as mice and rabbits, carrying around an ever-greater variety of viruses during an ever longer period of time. Perhaps we might say that man will be obligated to maintain them.

Viral density per species might well increase enormously. Recombination of viruses that are forced to inhabit the same cell together will definitely lead to the generation of new viruses and new virus families. It would be a viral evolution at unprecedented speeds. We will only contract illnesses from these viruses if they have trouble maintaining themselves, but that will almost certainly be the case. So many species of viruses in accommodations which are too small will inevitably lead to aggression. There will be more and more people who are afflicted with viruses and who die from them as well. And all this because viruses are being cornered.

If we do not actively defend ourselves with vaccines against the virus invasion, the chances for survival of many people on several continents might very well become smaller than we care to see. All these developments may be considered natural developments that are heading our way and against which we can arm ourselves. Mass vaccination will have to become the cornerstone of health care more than ever before, definitely at a time when there is a risk that exterminated viruses such as smallpox will be circulated again, that pandemics will be caused by viruses with an originally limited area of distribution, such as Ebola, the Spanish flu, and West Nile or that new viruses, like the SARS virus, will emerge. These viruses are there to stay: a proof of the immense fitness of viruses.

GLOSSARY

AIDS Virus   The human immunodeficiency virus (HIV), which causes a breakdown of the immune system, enabling other viruses, bacteria and parasites to make people seriously ill.

Anthropocene   The period from the end of the 18th century until the present, during which man has come to play the dominant role in nature.

Archaea   One of the oldest, if not the oldest, form of life on earth. These prehistoric bacteria are resistant against high temperatures, low acidity (pH) and high pressure.

Avian Flu Viruses   Viruses belonging to the influenza family that infect mainly birds, sometimes causing disease in humans; examples of the last year's viruses are H5N1 in Hongkong and H7N7 in the Netherlands.

B Cells   Cells of that part of the immune system that produces antibodies. Virus-specific antibodies bind to the virus in the process of making it impossible for viruses to infect cells.

Bacteriophages   Viruses of bacteria consisting of, on the one hand, virus families that kill bacteria and, on the other, virus families that increase the chances of survival for bacteria.

Bioterrorism   The intentional spreading of microorganisms—viruses and bacteria—for the purpose of causing illness or death in people.

Burkitt's Lymphoma   A form of cancer that occurs mainly in Africa and is caused by a herpes virus, the Epstein-Barr Virus (EBV).

Circo-viruses   Very small viruses of mammals and birds which seem to be the offspring of plant viruses (nanoviruses).

Complex Retroviruses   Viruses that reproduce by making DNA from RNA, thus becoming retroviruses, and that need a number of extra genes—sometimes as many as six—in order to maintain themselves in a host. HIV, the AIDS virus, is an example of a complex retrovirus.

Contaminated Food   Food may be contaminated with a human virus and thus infect those who eat the food. Food may also be contaminated with a plant virus or an animal virus that deprives the agricultural or livestock products of their nutritional value, thus possibly leading to famine among people.

Co-receptors   A virus can only penetrate a cell if it attaches itself to molecules that are found on the cell surface. Such molecules are called virus receptors or co-receptors.

Coronaviruses   A group of viruses recognized by a "crown" in electron microscopic images and linked to common colds and diarrhea, most recently a new coronavirus was shown to cause SARS.

Dengue   A flavivirus that occurs mainly in Asia, specifically in the big cities. Dengue is transmitted by mosquitoes and survives in man by continually taking on different forms—antigenic variation—in order to evade the human immune system.

Ebola   A virus that causes a deadly hemorrhagic fever through direct contact. The virus occurs exclusively in the interior of Africa, but recently has been associated with possible attacks of bioterrorism.

El Niño   A warming of the surface water in the Southern Pacific Ocean, amounting to an increase in temperature of about one degree Celsius. Temperature changes in the surface water of the oceans have been linked to cholera outbreaks.

El Niño Southern Oscillation (ENSO)   Joint term for the Southern Oscillation (air pressure change) and the El Niño effect (change in temperature of ocean surface)

Encephalitis Lethargica   A form of sleeping sickness that is being linked directly with having contracted the flu of 1918. This serious outcome of the flu is also referred to as Economo's Disease.

Endogenous Retroviruses   These are retroviruses that go into hiding as DNA copies in their host's genetic material, thus becoming an inextricable part. These virus DNAs are subsequently transmitted from offspring to offspring of the host.

Entero Viruses   A group of viruses, such as the polio virus and the Coxsackie virus, that are spread when feces end up in drinking water or when food becomes contaminated after having been prepared by an unsanitary cook. The mouth is the most important route for infection.

Epstein-Barr Virus (EBV)   A herpes virus that causes Pfeiffer's disease and has been linked to Burkitt's lymphoma in Africa and nasopharyngeal carcinoma in Asia.

Equine Infectious Anemia Virus (EIAV)   A virus of horses closely resembling the human AIDS virus. This disease of horses caused by the virus was successfully eradicated in China by vaccination with a live weakened strain of EIAV.

Eradication   Elimination of an infectious disease through human intervention. The best example is smallpox, which was eliminated from the face of the earth through vaccination. Rinderpest and polio, and perhaps measles, are faced with a similar fate.

Fitness   A member of the same or another virus family is considered more fit than another member of the family when it reproduces and spreads more efficiently than the other, in short, when it has a competitive advantage and its survival changes in a given environment are increased.

Flaviviruses   This family of viruses derives its name from "flavus," meaning yellow, and thus from the yellow fever virus. These viruses spread generally by mosquitoes bites. The yellow fever virus, West Nile Virus, and dengue belong all to this virus family.

Gal Antigen and Gal Antibodies   Humans and anthropoid apes, chimpanzees, gorillas and orangutans, do not have the Gal antigen on their cells, while practically every other mammal does have it. Conversely, human and its close relatives have Gal antibodies, while they are lacking in practically every mammal. Gal antibodies protect us against attacks by simple retroviruses.

Hanta Virus   A mouse virus that strikes man during periods when infestations of mice take place. People are infected by breathing in air contaminated with mouse urine containing the virus.

Herpes Viruses   Virus family to which belong the herpes simplex virus (HSV type 1 and 2), the virus which causes Pfeiffer's disease (EBV), and the virus which leads to Kaposi's sarcoma (KSHV or HHV8).

Heterozygous Viruses   Viruses that carry two distinct nucleic acid strings, generally RNA, which are usually identical but not in this particular case.

Human Virus   A virus that has the human species as its only host and which is barely able or unable to reproduce in any other host.

Kaposi's Sarcoma   An abnormal growth of blood vessels which spread over the body as pinkish red spots. It is caused by the human herpes virus type 8 (HHV8) also called Kaposi's sarcoma herpes virus (KSHV).

Keshan's Disease   A serious heart disorder that mainly strikes young children and their mothers in China. It is caused by an entero virus, the Coxsackie B virus, in combination with a Selenium deficiency due to undernourishment.

Kuru   A prion disease occurring exclusively in the highlands of New Guinea, which is caused by the consumption of infected brains of fellow tribespeople.

Live Attenuated (or Weakened) Virus Vaccine   A form of vaccine given to people in order to infect them with a virus that is closely related to the virus against which protection must be given, but that causes no symptoms of illness, or only very mild ones. An infection with such a crippled virus protects against aggressive viruses of the same or related virus families.

Mad Cow Disease   An illness in cows caused by prion protein changed in conformation, which spreads among cows by feeding them diseased brains of their own species. It was transmitted to man through the consumption of contaminated meat.

Measles   A virus infection without an animal pool, that causes a typical skin rash and which has a serious outcome only in exceptional cases. After smallpox, rinderpest and polio, measles is the next virus infection on the list for eradication.

Monkey Pox   An illness of monkeys caused by a virus that is closely related to the human smallpox virus. This simian virus can cause illness and subsequent death in people.

Morbillization   Like variolization for smallpox, the first attempts to protect against measles by rubbing measles lesions out over the skin, a procedure that remained without success.

Mouse Pox   A very deadly illness caused by a genetically manipulated, originally harmless mouse pox virus.

Nanoviruses   See Circo-viruses

Nasopharyngeal Carcinoma   A cancer of the oro-pharyngeal cavity caused by the combination of an infection with the Epstein-Barr Virus (EBV) and the consumption of salted fish.

Nipah Virus   A virus related to measles, that is lethal for pigs and which has also struck pig slaughterers in Malaysia.

Nitrosamines   Substances that turn cells infected with the Epstein-Barr Virus (EBV) into cancer cells. Nitrosamines get into the oro-pharyngeal cavity together with salted fish, thus contributing to the contraction of nasopharyngeal carcinoma.

Original Antigenic Sin   A phenomenon that occurs mainly with the flu virus, whereby exposure to a new flu virus activates immunity to particular against flu viruses with which the person has come into contact in previous years or in his youth.

Papilloma Viruses   Viruses that are the cause of warts and of some forms of cancer, specifically cancer of the cervix in women.

Phage Therapy   Treatment of a bacterial infection with bacteriophages.

This treatment went out of fashion when antibiotics were discovered and it became popular again with the arrival of resistance against antibiotics.

Polio Disease caused by a virus that enters the body with food and water by way of the intestines. This paralysis is a rare manifestation of a polio virus infection that generally runs its course unnoticed.

Potato Viruses Plant viruses that preferably infect potatoes, and jeopardize harvests and, therefore, the food supply.

Prions Proteins naturally occurring in the body that have an abnormal structure that causes this form of protein to stimulate other normally structured prion proteins to accumulate, specifically in the brain.

Proviral DNA Retroviral DNA that is incorporated in a stable way in the DNA of the host.

Quasi-species A word invented by Manfred Eigen describing the behavior of large numbers of viruses that all differ slightly from each other and that, as a group, behave more or less like a species of living creatures behaves, but in a shorter timespan which shorten cycles of reproduction.

Rabbit Pox Myxomatosis, a disease that is lethal to rabbits and which is caused by a pox virus, belonging to the same family as smallpox and monkeypox.

Reassortant Virus A virus, such as a flu virus, borrowing some strings of its separate strings of genomic information from one virus and others from another virus.

Recombinant Virus A virus containing parts of one virus and parts of another virus within one single string of RNA, within one single gene or several genes.

Reverse Transcriptase An enzyme called "reverse transcriptase" which is able to convert RNA into DNA, thus behaving in a way opposite to enzymes which convert DNA into RNA. Retroviruses are dependent on this enzyme for their survival.

Rinderpest Virus This virus is the cause of a deadly pest among cows, sheep and goats. It is closely related to the measles virus.

Sabin Vaccine A polio vaccine based on weakened (attenuated) live strains of the virus. It is a form of a live attenuated (weakened) virus vaccine. The vaccine was named after Sabin, who discovered it.

Salk Vaccine A polio vaccine based on virus that has been killed. The vaccine was named after Salk, who discovered it.

SARS Severe Acute Respiratory Syndrome (SARS) is a disease caused by a new coronavirus that is more virulent than other human coronaviruses.

Simple Retroviruses Retroviruses that infect and sicken many species of animals, but not people. These viruses are called simple because they only have three basic genes.

Smallpox   Viral disease that always manifests itself in a pox-scarred face. In the past, one out of every three infected people would die. Since the end of the 1970s, smallpox no longer occurs.

Southern Oscillation   Fluctuations in air pressure between Darwin, Australia, and Tahiti.

Superinfection   A new infection of a cell or organism that has already been infected with the same species of virus. In a number of cases, the cell is protected by the first infection against the second infection (superinfection interference).

Susceptibility   Susceptibility to a virus may be determined genetically or by environmental factors. Viruses do not infect everyone, nor do they cause the same degree of illness in everybody. The seriousness of a measles infection is affected by vitamin A. Selenium, or rather a deficiency of selenium in the diet, determines the seriousness of certain entero virus infections.

Syncytin   The protein that causes fusion of placental cells and which is coded by the envelope gene of a retrovirus.

T Cells   Cells of that part of the immune system that stimulate B cells to become active or destroy virus-infected cells directly.

Vaccination   The process of vaccinating. The word was derived from bringing about contact with cow pox virus, or vaccinia, for the purpose of protecting against an attack by the smallpox virus.

Vaccine   A product that activates the human immune system against a specific virus, thus preventing infection or illness due to the virus. A vaccine prepares the defense system for the attack of a virus by posing as the virus involved.

Vaccinia   Cow pox virus, see Vaccine and Vaccination.

Variant Creutzfeldt-Jakob Disease   The incurable Creutzfeldt-Jakob disease is a prion disease that randomly strikes $1{:}10^6$ people and that occurs more often in some families. Variant Creutzfeldt-Jakob is the disease variant caused by consumption of mad-cow meat.

Variola   The smallpox virus, which can be distinguished in two variants: Variola minor, with a mortality of 1%, and Variola maior, with a mortality of 30%. Variola minor protects against Variola maior. See Vaccination and Variolization.

Variolization   The process of intentionally causing an infection with a relatively harmless smallpox virus, the variola minor, by pricking a smallpox blister, dipping a thread into the fluid and then placing it into a small cut in the skin, in order to protect against variola maior.

Viral Altruism   The phenomenon where a virus actively contributes to the host's survival or the survival of a virus from a non-related virus family.

Viral Genealogy   The family relationships of a virus as drawn up on the basis of a comparison of viral genes, characteristics of viral proteins, and the form or appearance of the viruses.

Viral Survival Viral survival is essentially best defined as the process through which a virus avoids dying out. A virus or virus family dies out when the virus can no longer be found in any individual of the host species involved. The smallpox virus is an example.

Viral Vectors Transmitters of the virus from one host or host species—see Virus Reservoir—to another. Mosquitoes are a good example.

Virulence The ability to cause disease and the deadliness of a virus. The survival of a virus, and thus its distribution, often depends on a certain degree of virulence.

Virus Viruses belong to a life form found in the peripheral area between life and death. Viruses can reproduce, but they can only do so if they are helped by the cell into which they penetrate.

Virus Density The variety of viruses and the number of virus particles that are present in the complete population of a specific host species.

Virus Families Groups of viruses that differ based on host specificity or ability to cause disease, although they resemble each other based on the comparison of viral genes, characteristics of viral proteins, and the form or appearance of the viruses.

Virus Reservoir A host species of the virus in which the virus goes into hiding without causing any external manifestations, that is, illness, and from which it re-emerges under specific circumstances, in order to threaten man or another species of animal.

West Nile Virus A virus belonging to the family of flaviviruses, as does the yellow fever virus. This avian virus is spread by mosquitoes and it is now spreading not only in Europe, the Middle East, and Africa, but also in America.

Xenotransplantation The replacement of human organs, such as kidneys, hearts and lungs, with organs of animals, such as baboons and pigs, which carries the risk of infection with baboon and hog viruses.

Zooplankton Minuscule marine animals living on bacteria.

BIBLIOGRAPHY

Chapter I

Claas E.C.J., A.D.M.E. Osterhaus, R. van Beek, J.C. de Jong, G.F. Rimmelzwaan, D.A. Senne, S. Krauss, K.F. Shortridge, and R.G. Webster. "Human influenza A H5N1 virus related to a highly pathogenic avian influenza virus." *The Lancet* 351: 472–77, 1998.

Crosby A.W. *America's forgotten pandemic: the influenza of 1918.* Cambridge: Cambridge University Press, 1989.

Davies P. *Catching cold: 1918's forgotten tragedy and the scientific hunt for the virus that caused it.* London: Michel Joseph, 1999.

Fitch W.M., R.M. Bush, C.A. Bender, and N.J. Cox. "Long term trends in the evolution of H(3) HA1 human influenza type A." *PNAS* 94: 7712–18, 1997.

Gill P.W. and A.M. Murphy. "Naturally acquired immunity to influenza type A: lessons from two coexisting subtypes." *Med J Aust* 142: 94–98, 1985.

Guan Y., K.F. Shortridge, S. Krauss, and R.G. Webster. "Molecular characterization of H9N2 influenza viruses: were they the donors of the "internal" genes of H5N1 viruses in Hong Kong?" *PNAS* 96: 9363–67, 1999.

Hatta M., P. Gao, P. Halfmann and Y. Kawaoka. "Molecular basis for high virulence of Hong Kong H5N1 influenza A viruses." *Science* 293: 1840–45, 2001.

Kolata G. *Flu: the story of the great influenza pandemic of 1918 and the search for the virus that caused it.* New York: Farrar, Straus and Giroux, 1999.

McCall S., J.M. Henry, A.H. Reid, and J.K. Taubenberger. "Influenza RNA not detected in archival brain tissues from acute encefalitis lethargica cases or in postencepalitic Parkinson cases." *J Neuropathol Exp Neurol* 60: 696–704, 2001.

Oxford J.S. "Influenza A pandemics of the 20th century with special reference to 1918: virology, pathology and epidemiology." *Rev Med Virol* 10: 119–33, 2000.

Peiris J.S.M., Y. Guan, D. Markwell, P. Ghose, R.G. Webster, and K.F. Shortridge. "Cocirculation of avian H9N2 and contemporary 'human' H3N2 influenza A viruses in pigs in Southeast China: potential for genetic reassortment?" *J Virol* 75: 9679–86, 2001.

Reid A.H., T.G. Fanning, J.V. Hultin, and J.K. Taubenberger. "Origin and evolution of the 1918 'Spanish' influenza virus hemagglutinin gene." *PNAS* 96: 1651–56, 1999.

Skripchenko G.S., E.M. Poliakov and N.I. Kniazeva. "Immunological precursors of influenza epidemics." *Zh Mikrobiol Epidemiol Immunobiol* 5: 70–77, 1978.

Taubenberger J.K., A.H. Reid, A.E. Krafft, K.E. Bijwaard, and T.G. Ganning. "Initial genetic characterization of the 1918 'Spanish' influenza virus." *Science* 275: 1793–96, 1997.

Chapter 2

Beck M.A. "Increased virulence of Coxsackie-virus B3 in mice due to vitamin E or selenium deficiency." *American Society for Nutritional Sciences*, symposium "Newly emerging viral diseases: what role for nutrition", 966S-70S, 1997.

Cohen J.E. *How many people can the earth support?* New York: W.W. Norton & Company, 1995.

Chezzi C., N.K. Blackburn, and B.D. Schoub. "Molecular epidemiology of type 1 polioviruses from Africa." *J Gen Virol* 78: 1017–24, 1997.

Clarke T. "Polio's last stand." *Nature* 409: 278–80, 2001.

Diener T.O. "Origin and evolution of viroids and viroid-like satellite RNAs." *Virus Genes* 11: 119–31, 1996.

Editorial. "Polio eradication: the endgame." *Nat. Med.* 7: 131, 2001.

Eggers H.J. "Milestones in early poliomyelitis research (1840 to 1949)." *J Virol* 73: 4533–35, 1999.

Feldstein P.A., Y. Hu, and R.A. Owens. "Precisely full length, circularizable, complementary RNA: an infectious form of potato spindle tuber viroid." *PNAS* 95: 6560–65, 1998.

Gibbs M.J. and G.F. Weiller. "Evidence that a plant virus switched hosts to infect a vertebrate and then combined with a vertebrate-infecting virus." *PNAS* 96: 8022–27, 1999.

Levander O.A. and M.A. Beck. "Interacting nutritional and infectious etiologies of Keshan disease." *Biol Trace Elem Res* 56: 5–21, 1997.

Liu H.M., D.P. Zheng, L.B. Zhang, M.S. Oberste, M.A. Pallansch, and O.M. Kew. "Molecular evolution of a type 1 wild-vaccine poliovirus recombinant during widespread circulation in China." *J Virol* 74: 11153–61, 2000.

Martín J., G.L. Ferguson, D.J. Wood, and P.D. Minor. "The vaccine origin of the 1968 epidemic of type 3 poliomyelitis in Poland." *Virol* 278: 42–49, 2000.

Martínez-Soriano J.P., J.G. Galindo-Alonso, C.J.M. Maroon, I. Yucell, D.R. Smith, and T.O. Diener. "Mexican papita viroid: putative ancestor of crop viroids." *PNAS* 93: 9397–9401, 1996.

Melnick J.L. "Current status of poliovirus infections." *Clin Microbiol Rev* 9: 293–300, 1996.

Yang C.F., T. Naguib, S.J. Yang, E. Nasr, J. Jorba, N. Ahmed, R. Campagnoli, H. van der Avoort, H. Shimizu, T. Yoneyama, T. Miyamura, M. Pallansch, and O. Kew. "Circulation of endemic type 2 vaccine-derived poliovirus in Egypt from 1983–1993." *J Virol* 77: 8366–8377, 2003.

Chapter 3

Andrews N.J., C.P. Farrington, H.J.T. Ward, S.N. Cousens, P.G. Smith, A.M. Molesworth, R.S.G. Knight, J.W. Ironside, and R.G. Will. "Deaths from variant Creutzfeldt-Jakob disease in the UK." *The Lancet* 361: 751–752, 2003.

Brown P., R.G. Will, R. Bradley, D.M. Asher, and L. Detwiler. "Bovine spongiform encephalopathy and variant Creutzfeldt-Jakob disease: background, evolution, and current concerns." *Emerg Infect Dis* 7: 6–14, 2001.

Chua K.B., W.J. Bellini, P.A. Rota, B.H. Harcourt, A. Tamin, S.K. Lam, T.G. Ksiazek, P.E. Rollin, S.R. Zaki, W.J. Shieh, C.S. Goldsmith, D.J. Gubler, J.T. Roehrig, B. Eaton, A.R. Gould, J Olson, H. Field, P. Daniels, A.E. Ling, C.J. Peters, L.J. Anderson, and B.W.J. Mahy. "Nipah virus: a recently emergent deadly paramyxovirus." *Science* 288: 1432–35, 2000.

Gajdusek D.C., C.J. Gibbs Jr, and M. Alpers. "Experimental transmission of a kuru-like syndrome in chimpanzees." *Nature* 209: 794–96, 1966.

Ghani A.C., N.M. Ferguson, C.A. Donnelly, and R.M. Anderson. "Predicted vCJD mortality in Great Britain." *Nature* 406: 583–84, 2000.

Gibbs C.J. Jr and D.C. Gajdusek. "Transmission of scrapie to the cynomolgus monkey (*Macaca fascicularis*)." *Nature* 236: 73–4, 1972.

Gibbs C.J. Jr, D.C. Gajdusek, D.M. Asher, M.P. Alpers, E. Beck, P.M. Daniel, and W.B. Matthews. "Creutzfeldt-Jakob disease (subacute spongiform encephalopathy): transmission to the chimpanzee." *Science* 161: 388–89, 1968.

Goodman S. "More funding needed to wipe out rinderpest." *Nature* 411: 403, 2001.

Krakauer D.C., M. Pagel, and T.R.E. Southwood. "Phylogenesis of prion protein." *Nature* 380: 675, 1996.

Lee H.S., P. Brown, L. Cervenáková, R.M. Garruto, M.P. Alpers, D.C. Gajdusek, and L.G. Goldfarb. "Increased susceptibility to Kuru of carriers of the PRNP 129 methionine/methionine genotype." *J Infect Dis* 183: 192–96, 2001.

Mackenzie J.S., K.B. Chua, P.W. Daniels, B.T. Eaton, H.E. Field, R.A. Hall, K. Halpin, C.A. Johansen, P.D. Kirkland, S.K. Lam, P. McMinn, D.J. Nisbet, R. Paru, A.T. Pyke, S.A. Ritchie, P. Siba, D.W. Smith, G.A. Smith, A.F. van den Hurk, L.F. Wang, and D.T. Williams. "Emerging viral diseases of Southeast Asia and the Western Pacific." *Emerg Infect Dis* 7: 497–504, 2001.

Mead S., M.P.H. Stumpf, J. Whitfield, J.A. Beck, M. Poulter, T. Campbell, J.B. Uphill, D. Goldstein, M. Alpers, E.M.C. Fisher, and J. Collinge. "Balancing selection at the prion protein gene consistent with prehistoric kurulike epidemics." *Science* 300: 640–643, 2003.

Parashar U.D., L.M. Sunn, F. Ong, A.W. Mounts, M.T. Arif, T.G. Ksiazek, M.A. Kamaluddin, A.N. Mustafa, H. Kaur, L.M. Ding, G. Othman, H.M. Radzi, P.T. Kitsutani, P.C. Stockton, J. Arokiasamy, H.E. Gary Jr, and L.J. Anderson. "Case-control study of risk factors for human infection with a new zoonotic paramyxovirus, Nipah virus, during a 1998–1999 outbreak of severe encefalitis in Malaysia." *J Infect Dis* 181: 1755–59, 2000.

Rhodes R. *Deadly feasts: tracking the secrets of a terrifying new plague.* New York: Simon and Schuster, 1997.

Schätzl H.M., F. Wopfner, S. Gilch, A. von Brunn and G. Jäger. "Is codon 129 of prion protein polymorphic in human beings but not in animals?" *The Lancet* 349: 1603–04, 1997.

Tierney P. "The fierce anthropologist." *The New Yorker,* 9 October: 50–61, 2000.

Tompa P., G.E. Tusnády, Cserzö and I. Simon. "Prion protein: evolution caught and route." *PNAS* 98: 4431–36, 2001.

Windl O., M. Dempster, J.P. Estibeiro, R. Lathe, R. de Silva, T. Esmonde, R. Will, A. Springbett, T.A. Campbell, K.C.L. Sidle, M.S. Palmer, and J. Collinge. "Genetic basis of Creutzfeldt-Jakob disease in the United Kingdom: a systematic analysis of predisposing mutations and allelic variation in the PRNP gene." *Hum Genet* 98: 259–64, 1996.

Windl O., A. Giese, W. Schulz-Schaeffer, I. Zerr, K. Skworc, S. Arendt, C. Oberdieck, M. Bodemer, S. Poser, and H.A. Kretzschmar. "Molecular genetics of human prion diseases in Germany." *Hum Genet* 105: 244–52, 1999.

Chapter 4

Barrow P.A. and J.S. Soothill. "Bacteriophage therapy and prophylaxis: rediscovery and renewed assessment of potential." *Trends Microbiol* 5: 268–17, 1997.

Bernhardt T.G., I.N. Wang, D.K. Struck, and R. Young. "A protein antibiotic in the phage Qß virion: diversity in lysis targets." *Science* 292: 2326–29, 2001.

Bratbak G. and M. Heldal. "Viruses rule the waves: the smallest and most abundant members of marine ecosystems." *Microbiol Today* 27: 171–73, 2000.

Cairns J., G.S. Stent, and J.D. Watson, eds. *Phage and the origins of Molecular Biology.* New York: Cold Spring Harbor Laboratory Press, 1992.

Chao L. "Fitness of RNA virus decreased by Muller's ratchet." *Nature* 348: 454–55, 1990.

Colwell R.R. "Global climate and infectious disease: the cholera paradigm." *Science* 274: 2025–31, 1996.

Fuhrman J.A. "Marine viruses and their biogeochemical and ecological effects." *Nature* 399: 541–48, 1999.

Kovats R.S., M.J. Bouma, S. Hajat, E. Worrall, and A. Haines. "El Niño and health." *The Lancet* 362: 4993–5001, 2003.

Lewis S. *Arrowsmith*. New York: Signet Classic. 1926.

Lin W., K.J. Fullner, R. Clayton, J.A. Sexton, M.B. Rogers, K.E. Calia, S.B. Calderwood, C. Fraser, and J.J. Mekalanos. "Identification of a vibrio cholerae RTX toxin gene cluster that is tightly linked to the cholera toxin prophage." *PNAS* 96: 1071–76, 1999.

Lobitz B., L. Beck, A. Huq, B. Wood, G. Fuchs, A.S.G. Faruque, and R. Colwell. "Climate and infectious disease: use of remote sensing for detection of vibrio cholerae by indirect measurement." *PNAS* 97: 1438–43, 2000.

Lucchini S., F. Desiere, and H. Brüssow. "Comparative genomics of streptococcus thermophilus phage species supports a modular evolution theory." *J Virol* 73: 8647–56, 1999.

Nowak M.A. and K. Sigmund. "Phage-lift for game theory." *Nature* 398: 367–68, 1999.

Richardson S.H. and D.J. Wozniak. "An ace up the sleeve of the cholera bacterium." *Nat Med* 2: 853–55, 1996.

Ruiz G.M., T.K. Rawlings, F.C. Dobbs, L.A. Drake, T. Mullady, A. Huq, and R.R. Colwell. "Global spread of microorganisms by ships." *Nature* 408: 49, 2000.

Summers W.C. *Félix d'Herelle and the origins of molecular biology*. New Haven, CT: Yale University Press, 1999.

Turner P.E. and L. Chao. "Prisoner's dilemma in an RNA virus." *Nature* 398: 441–43, 1999.

Waldor M.K. and J.J. Mekalanos. "Lysogenic conversion by a filamentous phageencoding cholera toxin." *Science* 272: 1910–14, 1996.

Weiner J. *Time, love, memory: a great biologist and his quest for the origins of behavior*. New York: Alfred A. Knopf, 1999.

Wommack K.E. and R.R. Colwell. "Virioplankton: viruses in aquatic ecosystems." *Microbiol Mol Biol Rev* 64: 69–114, 2000.

Chapter 5

Bonn D. "Hantaviruses: an emerging threat to human health?" *The Lancet* 352: 886, 1998.

Epstein P.R. "Climate and health." *Science* 285: 347–48, 1999.

Harper D.R. and A.S. Meyer. *Of mice, men and microbes: Hantavirus*. London: Academic Press, 1999.

Iwamoto M., D.B. Jerniganm A. Guasch, M.J. Trepka, C.G. Blackmore, W.C. Hellinger, S.M. Pham, S. Zaki, R.S Lanciotti, S.E. Lance-Parker, C.A. Diaz-Granados, A.G. Winquist, C.A. Perlino, S. Wiersma, K.L. Hillyer, J.L. Goodman, A.A. Marfin, M.E. Chamberland, and L.R. Petersen. "Transmission of West Nile Virus from an organ donor to four transpland recipients." *N Engl J Med* 348: 2196–2203, 2003.

Kuno G., G.J.J. Chang, K.R. Tsuchiya, N. Karabatsos, and C.B. Cropp. "Phylogeny of the genus flavivirus." *J Virol* 72: 73–83, 1998.

Lanciotti R.S., J.T. Roehrig, V. Deubel, J. Smith, M. Parker, K. Steele, B. Crise, K.E. Volpe, M.B. Crabtree, J.H. Scherret, R.A. Hall, J.S. MacKenzie, C.B. Cropp, B. Panigraphy, E. Ostlund, B. Schmitt, M. Malkinson, C. Banet, J. Weissmann, N. Komar, H.M. Savage, W. Stone, T. McNamara, and D.J. Gubler. "Origin of the West Nile Virus responsible for an outbreak of encephalitis in the Northeast United States." *Science* 286: 2333–37, 1999.

Linthicum K.J., A. Anyamba, C.J. Tucker, P.W. Kelley, M.F. Myers, and C.J. Peters. "Climate and satellite indicators to forecast Rift Valley fever epidemics in Kenya." *Science* 285: 397–400, 1999.

Morse D.L. "West Nile Virus – Not a passing phenomenon." *N Engl J Med* 348: 2173–2174, 2003.

Rappole J.H., S.R. Derrickson, and Z. Habálek. "Migratory birds and spread of West Nile Virus in the Western hemisphere." *Emerg Infect Dis* 6: 319–28, 2000.

Rigau-Pérez J.G., G.G. Clark, D.J. Gubler, P. Reiter, E.J. Sanders and A. Vance Vorndam. "Dengue and dengue haemorrhagic fever." *The Lancet* 352: 971–77, 1998.

Rodríguez L.L., W.M. Fitch, and S.T. Nichol. "Ecological factors rather than temporal factors dominate the evolution of vesicular stomatitis virus." *PNAS* 93: 13030–35, 1996.

Zanotto P.M. de, E.A. Gould, G.F. Gao, P.H. Harvey, and E.C. Holmes. "Population dynamics of flaviviruses revealed by molecular phylogenies." *PNAS* 93: 548–53, 1996.

Chapter 6

Conrad B, R.N. Weissmahr, J. Böni, R. Arcari, J. Schüpbach, and B. Mach. "A human endogenous retroviral superantigen as candidate autoimmune gene in type 1 diabetes." *Cell* 90: 303–13, 1997.

Galili U. and K. Swanson. "Gene sequences suggest inactivation of ?-1,3–galactosyltransferase in catharrhines after the divergence of apes from monkeys." *PNAS* 88: 7401–04, 1991.

Galili U., R.E. Mandrell, R.M. Hamadeh, S.B. Shohet, and J.M. Griffiss. "Interaction between human natural anti-a-galactosyl immunoglobulin G and bacteria of the human flora." *Infect Immun* 56: 1730–37, 1988.

Galili U., M.R. Clark, S.B. Shohet, J. Buehler, and B.A. Macher. "Evolutionary relationship between the natural anti-Gal antibody and the Galα1–3Gal epitope in primates." *PNAS* 84: 1369–73, 1987.

Herniou E., J. Martin, K. Miller, J. Cook, M. Wilkinson, and M. Tristem. "Retroviral diversity and distribution in vertebrates." *J Virol* 72: 5955–66, 1998.

Johnson W.E. and J.M. Coffin. "Constructing primate phylogenies from ancient retrovirus sequences." *PNAS* 96: 10254–60, 1999.

Klenerman P., H. Hengartner, and R.M. Zinkernael. "A non-retroviral RNA virus persists in DNA form." *Nature* 390: 298–301, 1997.

Kuyl A.C. van der, J.T. Dekker, and J. Goudsmit. "Distribution of baboon endogenous virus among species of African monkeys suggests multiple ancient cross-species transmissions in shared habitats." *J Virol* 69: 7877–87, 1995.

Kuyl A.C. van der, R. Mang, J.T. Dekker, and J. Goudsmit. "Complete nucleotide sequence of simian endogenous type D retrovirus with intact genome organization: evidence for ancestry to simian retrovirus and baboon endogenous virus." *J Virol* 71: 3666–76, 1997.

Mang R., J. Maas, A.C. van der Kuyl, and J. Goudsmit. "Papio cynocephalus endogenous retrovirus among old world monkeys: evidence for coevolution and ancient cross-species transmissions." *J Virol* 74: 1578–86, 2000.

Mang R., J. Goudsmit, and A.C. van der Kuyl. "Novel endogenous type C retrovirus in baboons: complete sequence, providing evidence for baboon endogenous virus gag-pol ancestry." *J Virol* 73: 7021–26, 1999.

Mi S., X. Lee, X. Li, G.M. Veldman, H. Finnerty, L. Racie, E. LaVallie, X.Y. Tang, P. Edouard, S. Howes, J.C. Keith Jr, and J.M. McCoy. "Syncytin is a captive retroviral envelope protein involved in human placental morphogenesis." *Nature* 403: 785–89, 2000.

Rother R.P., W.L. Fofor, J.P. Springhorn, C.W. Birks, E. Setter, M.S. Sandrin, S.P. Squinto, and S.A. Rollins. "A novel mechanism of retrovirus inactivation in human serum mediated by anti-a-galactosyl natural antibody." *J Exp Med* 182: 1345–55, 1995.

Schulte A.M. and A. Wellstein. "Structure and phylogenetic analysis of an endogenous retrovirus inserted into the human growth factor gene pleiotrophin." *J Virol* 72: 6065–72, 1998.

Takeuchi Y., C.D. Porter, K.M. Strahan, A.F. Preece, K. Gustafsson, F.L. Cosset, R.A. Weiss, and M.K.L. Collins. "Sensitization of cells and retroviruses to human serum by (a1–3) galactosyltransferase." *Nature* 379: 85–88, 1996.

Weinberg R.A. *Racing to the beginning of the road: the search for the origin of cancer.* New York: Harmony Books, 1996.

Weiss R.A. and P. Kellam. "Illicit viral DNA." *Nature* 390: 235–36, 1997.

Zhdanov V.M. "Integration of viral genomes." *Nature* 256: 471–73, 1975.

Chapter 7

Brown F. and A.M. Lewis Jr, eds. *Simian Virus 40 (SV40): a possible human polyomavirus. Developments in biological standardization,* dl. 94. Basel: Karger, 1998.

Cornelissen M., A.C. van der Kuyl, R. Van den Burg, F. Zorgdrager, C.J.M. van Noesel and J. Goudsmit. "Gene expression profile of AIDS-related Kaposi's sarcoma." *BMC Cancer* 3: 7, 2003.

Davidovici B., I. Karakis, D. Bourboulia, S. Ariad, J.C. Zong, D. Benharroch, N.

Dupin, R. Weiss, G. Hayward, B. Sarov, and C. Boshoff. "Seroepidemiology and molecular epidemiology of Kaposi's sarcoma-associated herpesvirus among Jewish population groups in Israel." *J Natl Cancer Inst* 93: 194–202, 2001.

Hatwell J.N. and P.M. Sharp. "Evolution of human polyomavirus JC." *J Gen Virol* 81: 1191–1200, 2000.

Imperiale M.J. "The human polyomaviruses, BKV and JCV: molecular pathogenesis of acute disease and potential role in cancer." *Virol* 267: 1–7, 2000.

Lacoste V., P. Mauclère, G. Dubreuil, J. Lewis, M.C. Georges-Courbot, and A. Gessain. "KSHV-like herpesviruses in chimps and gorillas." *Nature* 407: 151–52, 2000.

McGeoch D.J., A. Dolan and A.C. Ralph. "Toward a comprehensive phylogeny for mammalian and avian herpesviruses." *J Virol* 74: 10401–06, 2000.

Ong C.K., S. Nee, A. Rambaut, H.U. Bernard, and P.H. Harvey. 'Elucidating the population histories and transmission dynamics of papillomaviruses using phylogenetic trees." *J Mol Evol* 44: 199–206, 1997.

Renwick N., T. Halaby, G.J. Weverling, N.H.T.M. Dukers, G.R. Simpson, R.A. Coutinho, J.M.A. Lange, T.F. Schulz, and J. Goudsmit. "Seroconversion for human herpesvirus 8 during HIV infection is highly predictive of Kaposi's sarcoma." *AIDS* 12: 2481–2488, 1998.

Sarid R., S.J. Olsen, and P.S. Moore. "Kaposi's sarcoma-associated herpesvirus: epidemiology, virology, and molecular biology." In: *Advances in virus research*, dl 52, 139–232. Eds. K. Maramorosch, F.A. Murphy, and A.J. Shatkin. New York: Academic Press, 1999.

Sugimoto C., T. Kitamura, J. Guo, M.N. Al-Ahdal, S.N. Shchelkunov, B. Otova, P. Ondrejka, J.Y. Chollet, S. El-Safi, M. Ettayebi, G. Grésenguet, T. Kocagöz, S. Chaiyarasamee, K.Z. Thant, S. Thein, K. Moe, N. Kobayashi, F. Taguchi, and Y. Yogo. "Typing of urinary JC virus DNA offers a novel means of tracing human migrations." *PNAS* 94: 9191–96, 1997.

Vanderheijden N, L.P. Delputte, H.W. Favoreel, J. Vandekerckhove, J. van Damme, P.A. van Woensel, and H.J. Nauwynck. "Involvement of sialoadhesin in entry of porcine reproductive and respiratory syndrome virus into porcine alveolar macrophages". *J. Virol* 77: 8207–8215, 2003.

Zou X.N., S.H. Lu, and B. Liu. "Volatile N-nitrosamines and their precursors in Chinese salted fish: a possible etiological factor for NPC in China." *Int J Cancer* 59: 155–58, 1994.

Chapter 8

Blancou P., J.P. Vartanian, C. Christopherson, N. Chenciner, C. Bassillico, S. Kwok, and S. Wain-Hobson. "Polio vaccine samples not linked to AIDS." *Nature* 410: 1045–46, 2001.

Blower S.M., K. Koelle, D.E. Kirschner and J. Mills. "Live attenuated HIV vaccines: predicting the tradeoff between efficacy and safety." *PNAS* 98: 3618–23, 2001.

Chadwick B.J., M. Desport, J. Brownlie, G.E. Wilcox, and D.M.N. Dharma. "Detection of Jembrana disease virus in spleen, lymph nodes, bone marrow and other tissues by in situ hybridization of paraffin-embedded sections." *J Gen Virol* 79: 101–6, 1998.

Chitnis A., D. Rawls, and J. Moore. "Origin of HIV type 1 in colonial French Equatorial Africa?" *AIDS Res Hum Retrov* 16: 5–8, 2000.

Gao F., E. Bailes, D.L. Robertson, Y. Chen, C.M. Rodenburg, S.F. Michael, L.B. Cummins, L.O. Arthur, M. Peeters, G.M. Shaw, P.M. Sharp, and B.H. Hahn. "Origin of HIV-1 in the chimpanzee Pan troglodytes." *Nature* 397: 436–40, 1999.

Goldstein S., I. Ourmanov, C.R. Brown, B.E. Beer, W.R. Elkins, R. Plishka, A. Buckler-White, and V.M. Hirsch. "Wide range of viral load in healthy African Green Monkeys naturally infected with simian immunodeficiency virus." *J Virol* 74: 11744–53, 2000.

Goudsmit J. *Confidential report on EIA donkey leucocyte- attenuated virus vaccine passage 124 and 125 of the Harbin Veterinary Research Institute, China: preliminary evaluation of protection rates and side effects in donkeys and horses.* 1997.

Goudsmit J. *Viral sex: the nature of AIDS.* New York: Oxford University Press, 1997.

Goudsmit J. and V.V. Lukashov. "Dating the origin of HIV-1 subtypes." *Nature* 400: 325–26, 1999.

Hahn B.H., G.M. Shaw, K.M. de Cock and P.M. Sharp. "AIDS as a zoonosis: scientific and public health implications." *Science* 287: 607–14, 2000.

Hillis D.M. "Origins of HIV." *Science* 288: 1757–59, 2000.

Korber B., M. Muldoon, J. Theiler, F. Gao, R. Gupta, A. Lapedes, B.H. Hahn, S. Wolinski, and T. Bhattacharya. "Timing the ancestor of the HIV-1 pandemic strains." *Science* 288: 1789–96, 2000.

Lemey P., O.G. Pybus, B. Wang, N.K. Saksena, M. Salemi, and A.M. Vandamme. "Tracing the origin and history of the HIV-2 epidemic." *PNAS* 100: 6588–6592, 2003.

Libert F., P. Cochaux, G. Beckman, M. Samson, M. Aksenova, A. Cao, Czeizel, M. Claustres, C. de la Rúa, M. Ferrari, C. Ferrec, G. Glover, B. Grinde, S.S.S. Güran, V. Kucinskas, J. Lavinha, B. Mercier, G. Ogur, L. Peltonen, C. Rosatelli, M. Schwartz, V. Spitsyn, L. Timar, L. Beckman, M. Parmentier, and G. Vassart. "The Δccr5 mutation conferring protection against HIV-1 in Caucasian populations has a single and recent origin in Northeastern Europe." *Hum Mol Genet* 7: 399–406, 1998.

Lucotte G., G. Mercier and P. Smets. "Elevated frequencies of the mutant allele Δ32 at the ccr5 gene in Asjkenazic jews." *AIDS Res Hum Retrov* 15: 1–2, 1999.

Martinson J.J., N.H. Chapman, D.C. Rees, Y.T. Liu, and J.B. Clegg. "Global distribution of the ccr5 gene 32–basepair deletion." *Nat Genet* 16: 100–3, 1997.

Myers G. "Tenth anniversary perspectives on AIDS: HIV: between past and future." *AIDS Res Hum Retrov* 10: 1317–24, 1994.

O'Brien S.J. and J.P. Moore. "The effect of genetic variation in chemokines and their receptors on HIV transmission and progression to AIDS." *Immunol Rev* 177: 99–111, 2000.

O'Brien S.J and M. Dean. "In search of AIDS-resistance genes." *Sci Am* 277: 44–51, 1997.

Proceedings of the International Symposium on Immunity to Equine Infectious Anemia. Harbin Veterinary Research Institute of Chinese Academy of Agricultural Sciences, 1983.

Roda Husman A.M. de, M. Koot, M.T.E. Cornelissen, I.P.M. Keet, M. Brouwer, S.M. Broersen, M. Bakker, M.T.L. Roos, M. Prins, F. de Wolf, R. Coutinho, F. Miedema, J. Goudsmit, and H. Schuitemaker. "Association between CCR5 genotype and the clinical course of HIV-1 infection." *Ann Intern Med* 127: 882–90, 1997.

Salemi M., K. Strimmer, W.W. Hall, M. Duffy, E. Delaporte, S. Mboup, M. Peeters, and A.M. Vandamme. "Dating the common ancestor of SIVcpz andHIV-1 group M and the origin of HIV-1 subtypes by using a new method to uncover clock-like molecular evolution." *FASEB J* 15: 276–78, 2001.

Smith M.W., M. Dean, M. Carrington, C. Winkler, G.A. Huttley, D.A. Lomb, J.J. Goedert, T.R. O'Brien, L.P. Jacobson, R. Kaslow, S. Buchbinder, E. Vittinghoff, D. Vlahov, K. Hoots, M.W. Hilgartner, Hemophilia Growth and Develoment Study (HGDS), Multicenter AIDS Cohort Studies (MACS), Multicenter Hemophilia Cohort Study (MHCS), San Francisco City Cohort (SFCC), ALIVE Study, and S.J. O'Brien. "Contrasting genetic influence of ccr2 and ccr5 variants on HIV-1 infection and disease progression." *Science* 277: 959–65, 1997.

Stephens J.C., D.E. Reich, D.B. Goldstein, H. Doo Shin, M.W. Smith, M. Carrington, C. Winkler, G.A. Huttley, R. Allikmets, L. Schriml, B. Gerrard, M. Lalasky, M.D. Ramos, S. Morlot, M. Tzetis, Oddoux, F.S. di Giovine, G. Nasioulas, D. Chandler, M. Aseev, M. Hanson, L. Kalaydjieva, D. Glavac, P. Gasparini, E. Kanavakis, M. Claustres, M. Kambouris, H. Ostrer, G. Duff, V. Baranov, H. Sibul, A. Metspalu, D. Goldman, N. Martin, D. Duffy, J. Schmidtke, X. Estivill, S.J. O'Brien, and M. Dean. "Dating the origin of ccr5-Δ32 AIDS-resistance allele by the coalescence of haplotypes." *Am J Hum Genet* 62: 1507–15, 1998.

Temin H.M. "A proposal for a new approach to a preventive vaccine against human immunodificiency virus type 1." *PNAS* 90: 4419–20, 1993.

Vangroenweghe D. *AIDS in Afrika*. Breda: De Geus, 1997.

Zhu T., B.T. Korber, A.J. Nahmiasi, E. Hooper, P.M. Sharp and D.D. Ho. "An African HIV-1 sequence from 1959 and implications for the origin of the epidemic." *Nature* 391: 594–97, 1998.

Chapter 9

Bazin H. *The eradication of smallpox: Edward Jenner and the first and only eradication of a human infectious disease*. London: Academic Press, 2000.

Best S.M. and P.J. Kerr. "Coevolution of host and virus: the pathogenesis of virulent and attenuated strains of Myxoma virus in resistant and susceptible European rabbits." *Virology* 267: 36–48, 2000.

Damaso C.R.A., J.J. Esposito, R.C. Condit, and N. Moussatché. "An emergent poxvirus from humans and cattle in Rio de Janeiro state: Cantagalo virus may derive from Brazilian smallpox vaccine." *Virology* 277: 439–49, 2000.

Department of Health and Human Services, Centers for Disease Control and Prevention "Update: multistate outbreak of monkeypox – Illinois, Indiana, Kansas, Missouri, Ohio and Wisconsin, 2003." *MMWR* 52: 642–646, 2003.

Dukers N.H., N. Renwick, M. Prins, R.B. Geskus, T.F. Schulz, G.J. Weverling, R.A. Coutinho, and J. Goudsmit. "Risk factors for human herpesvirus 8 seropositivity and seroconversion in a cohort of homosexual men." *Am J Epidemiol* 151: 213–24, 2000.

Evans A.S. and R.A. Kaslow, eds. *Viral infections of humans: epidemiology and control.* New York: Plenum Medical Book Company, 1997.

Fenner F. and B. Fantini. *Biological control of vertebrate pests: the history of myxomatosis, an experiment in evolution.* London: CABI Publishing, 1999.

Hutin Y.J.F., R.J. Williams, P. Malfait, R. Pebpdy, V.N. Loparev, S.L. Ropp, M. Rodriguez, J.C. Knight, F.K. Tshioko, A.S. Khan, M.V. Szczeniowski, and J.J. Esposito. "Outbreak of human monkeypox, Democratic Republic of Congo, 1996–1997." *Emerg Infect Dis* 7: 434–438, 2001.

Jackson R.J., A.J. Ramsay, C.D. Christensen, S. Beaton, D.F. Hall, and I.A. Ramshaw. "Expression of mouse interleukin-4 by a recombinant ectromelia virus suppresses cytolytic lymphocyte responses and overcomes genetic resistance to mousepox." *J Virol* 75: 1205–10, 2001.

Krings M., A. Stone, R.W. Schmitz, H. Krainitzki, M. Stoneking, and S. Pääbo. "Neandertal DNA sequences and the origin of modern humans." *Cell* 90: 19–30, 1997.

Lalani A.S. J. Masters, W. Zeng, J. Barrett, R. Pannu, H. Everett, C.W. Arendt, and G. McFadden. "Use of chemokine receptors by poxviruses." *Science* 286: 1968–1971, 1999.

Marzio G., K. Verhoef, M. Vink and B. Berkhout. "In vitro evolution of a highly replicating, doxycycline-depend HIV for applications in vaccine studies." *PNAS* 98: 6342–6347, 2001.

Mukinda V.B.K., G. Mwema, M. Kilundu, D.L. Heymann, A.S. Khan, J.J. Esposito, and other members of the Monkeypox Epidemiologic Working Group. "Re-emergence of human monkeypox in Zaire in 1996." *The Lancet* 349: 1449–50, 1997.

Ovchinnikov I.V., A. Gotherstrom, G.P. Romanova, V.M. Kharitonov, K. Liden, and W. Goodwin. "Molecular analysis of Neanderthal DNA from the northern Caucasus." *Nature* 404: 490–93, 2000.

Renwick N., T. Halaby, G.J. Weverling, N.H.T.M. Dukers, G.R. Simpson, R.A. Coutinho, J.M.A. Lange, T.F. Schulz, and J. Goudsmit. "Seroconversion for human herpesvirus 8 during HIV infection is highly predictive of Kaposi's sarcoma." *AIDS* 12: 2481–88, 1998.

Shisler J.L., T.G. Senkevich, M.J. Berry, and B. Moss. "Ultraviolet-induced cell death blocked by a selenoprotein from a human dermatotropic poxvirus." *Science* 279: 102–5, 1998.

Chapter 10

Burke D.S. "Six months to act; the SARS epidemic (opinion)." *Wall Street Journal*, April 25, 2003.

Cyranoski, D. and A. Abbott. "Virus detectives seek source of SARS in China's wild animals." *Nature* 423: 467, 2003.

Donnelly C.A., A.C. Ghani, G.M. Leung, A.J. Hedley, C. Fraser, S. Riley, L.J. Abu-Raddad, L.M. Ho, T.Q. Thach, P. Chau, K.P. Chan, T.H. Lam, L.Y. Tse, T. Tsang, S.H. Liu, J.H.B. Kong, E.M.C. Lau, N.M. Ferguson, and R.M. Anderson. "Epidemiological determinants of spread of causal agent of Severe Acute Respiratory Syndrome in Hong Kong." *The Lancet* 361: 1761–1766, 2003.

Drosten C., S. Günther, W. Preiser, S. van der Werf, H.R. Brodt, S. Becker, H. Rabenau, M. Panning, L. Kolesnikova, R.A.M. Fouchier, A. Berger, A.M. Burguière, J. Cinatl, M. Eickmann, N. Escriou, K. Grywna, S. Kramme, J.C. Manuguerra, S. Müller, V. Rickerts, M. Stürmer, S. Vieth, H.D. Klenk, A.D.M.E. Osterhaus, H. Schmitz, and H.W. Doerr. "Identification of a novel coronavirus in patients with Severe Acute Respiratory Syndrome." *N Engl J Med* 348: 1967–1976, 2003.

Fouchier R.A.M., T. Kuiken, M. Schutten, G. van Amerongen, G.J.J. van Doornum, B.G. van den Hoogen, M. Peiris, W. Lim, K. Stöhr, and A.D.M.E. Osterhaus. "Koch's postulates fulfilled for SARS virus." *Nature* 423: 240, 2003.

Ksiazek T.G., D. Erdman, C.S. Goldsmith, S.R. Zaki, Peret T., S. Emery, S. Tong, C. Urbani, J.A. Comer, W. Lim, P.E. Rollin, S.F. Dowell, A.E. Ling, C.D. Humphrey, W.J. Shieh, J. Guarner, C.D. Paddock, P. Rota, B. Fields, J. DeRisi, J.Y. Yang, J.M. Hughes, J.W. LeDuc, J.W. Bellini, L.J. Anderson, and the SARS Working Group. "A novel coronavirus associated with Severe Acute Respiratory Syndrome." *N Engl J Med* 348: 1953–1966, 2003.

Marra M.A., S.J.M. Jones, C.R. Astell, R.A. Holt, A. Brooks-Wilson, Y.S.N. Butterfield, J. Khattra, J.K. Asano, S.A. Barber, S.Y Chan, A. Cloutier, S.M. Coughlin, D. Freeman, N. Girn, O.L. Griffith, S.R. Leach, M. Mayo, H. McDonald, S.B. Montgomery, P.K. Pandoh, A.C. Petrescu, A.G. Robertson, J.E. Schein, A. Siddiqui, D.E. Smailus, J.M. Stott, and G.S. Yang. "The genome sequence of the SARS-associated coronavirus." *Science* 300: 1399–1404, 2003.

Mortimer P.P. "Five postulates for resolving outbreaks of infectious disease." *J Med Microbiol* 52: 447–451, 2003.

Peiris J.S.M., S.T. Lai, L.L.M. Poon, Y. Guan, L.Y.C. Yam, W. Lim, J. Nicholls, W.K.S. Yee, W.W. Yan, M.T. Cheung, V.C.C. Cheng, K.H. Chan, D.N.C. Tsang, R.W.H. Yung, T.K. Ng, K.Y. Yuen, and members of the SARS Study Group. "Coronavirus as a possible cause of Severe Acute Respiratory Syndrome". *The Lancet* 361: 1319–1325, 2003.

Rota P.A., M.S. Oberste, S.S. Monroe, W.A. Nix, R. Campagnoli, J.P. Icenogle, S. Peñaranda, B. Bankamp, K. Maher, M. Chen, S. Tong, A. Tamin, L. Lowe, M. Frace, J.L. DeRisi, Q. Chen, D. Wang, D.D. Erdman, T.C.T. Peret, C. Burns, T.G. Ksiazek, P.E. Rollin, A. Sanchez, S. Liffick, B. Holloway, J. Limor, K. Mc-Caustland, M. Olsen-Rassmussen, R. Fouchier, S. Günther, A.D.M.E. Osterhaus, C. Drosten, M.A. Pallansch, L.J. Anderson, and W.J. Bellini. "Characterization of a novel coronavirus associated with Severe Acute Respiratory Syndrome." *Science* 300: 1394–1399, 2003.

Ruan Y., C.L. Wei, L.A. Ee, V.B. Vega, H. Thoreau, S.T.S. Yun, J.M. Chia, P. Ng, K.P. Chiu, L. Lim, Z. Tao, C.K. Peng, L.O.L. Ean, N.M. Lee, L.Y. Sin, L.F.P. Ng, R.E. Chee, L.W. Stanton, P.M. Long, and E.T. Liu. "Comparative full-length genome sequence analysis of 14 SARS coronavirus isolates and common mutations associated with putative origins of infection." *The Lancet* 361: 1779–1784, 2003.

WHO Multicentre Collaborative Network for SARS Diagnosis. "A multicentre collaboration to investigate the cause of Severe Acute Respiratory Syndrome." *The Lancet* 361: 1730–1733, 2003.

Epilogue

Baranowski E., C.M. Ruiz-Jarabo, and E. Domingo. "Evolution of cell recognition by viruses." *Science* 292: 1102–05, 2001.

Crutzen P.J. "Geology of mankind." *Nature* 415: 23, 2002.

Diamond J. *Guns, germs and steel: a history of everybody for the last 13,000 years.* London: Vintage, 1998.

Domingo E., C. Escarmís, L. Menéndez-Arias and J.J. Holland. "Viral quasi-species and fitness variations." In: *Origin and evolution of viruses*, 141–61. Ed. Domingo et al. New York: Academic Press, 1999.

Griffiths P.D. "Herpesviruses: from the cradle to the grave." *Microbiol Today* 28: 182–84, 2001.

Hendrix R.W., J.G. Lawrence, G.F. Hatfull and S. Casjens. "The origins and ongoing evolution of viruses." *Trends Microbiol* 8: 504–08, 2000.

Jeffares D.C., A.M. Poole and D. Penny. "Relics from the RNA world." *J Mol Evol* 46: 18–36, 1998.

Knipe D.M. and P.M. Howley, eds. *Fields Virology*. Vols. 1 and 2. Lippincott Williams & Wilkins, 2001.

Mims, C. *The war within us: everyman's guide to infection and immunity*. London: Academic Press, 2000.

Poole A.M., D.C. Jeffares and D. Penny. "The path from the RNA world." *J Mol Evol* 46: 1–17, 1998.

Sáiz J.C. and E. Domingo. "Virulence as a positive trait in viral persistence." *J Virol* 70: 6410–13, 1996.

INDEX

access to antivirals, 119, 120
aggression and virus spread, 75
AIDS, African origin of, 113
AIDS resistance, 112
    origin of, 112
    related to smallpox, 112, 139
AIDS vaccines, 118–123
    equine AIDS vaccine, 122
    live-attenuated SIV, 121, 122
air and viruses. *See also* influenza
antibody treatment of viruses, 143
antiviral therapy against AIDS, 118,
    119
archaea, 4, 126
avian flu virus, 13
    Asian origins of, 23, 144, 145
    ducks and, 13
    epidemics of, 23
    infection of intestines, 13
    water spread of, 15

Bacillus comma. *See* cholera
Bacteria, 4
Bacteriophages, 56
    bacterial survival, 57
    interaction with bacteria, 57–58
    seawater and, 57

Baltimore, David, 79, 123
Bioterrorism, 134
    Ebola, 77
    smallpox, 134
bird flu. *See* avian flu virus
breakdown of immune system by
    HIV, 110
breast cancer, and viruses, 81, 82
BSE. *See* mad cow disease

catalytic RNA, 32
cervical cancer and virus, 104
cholera, 58–61
    and mussels and oysters, 63
    viruses of, 58
    and zooplankton, 63–65
climate and viruses. *See* weather and
    viruses
close contact and virus infection, 141
co-evolution of virus and host,
    104–106
Creutzfeldt-Jakob Disease, 49
    variant CJD, 49, 50
Crutzen, Paul, 153

defense against viruses, 109
diaspora of hosts and viruses, 138

disease caused by a virus. *See* Koch's postulates
drug resistance of viruses, 85, 86

Ebola, 77
El Niño Southern Oscillation (ENSO), 64, 65
    and mosquitoes, 76
    and virus spread, 64
encephalitis lethargica, and influenza, 18
enteroviruses, 34–36
eukarya, 4
evolution of viruses, 85

female sex and viruses, 103–105
Flaviviruses, 67
    and mosquitoes, 67
flu virus. *See also* influenza virus
food and viruses. *See also* poliovirus; enteroviruses;mad cow disease
food poisoning, and viruses, 34
foot-and-mouth disease, 42

Gajdusek, D. Carleton, 49

Hantavirus, 72–75
    aggression and, 75
    rodent origin of, 72, 74
d'Herelle, Felix, 56
herpes, 94–96
    and saliva, 95
    and sex, 95, 96
HIV, primate origin of, 113–116
hog flu, epidemic in 1976, 22
human evolution, 129

inactivated polio vaccine (IPV), 40
influenza virus,
    1918 epidemic of, 16
    evolution of, 24
    origin of, 14, 18
    and pigs, 14
    virulence of, 16
and young children, 16

jungle and virus origin, 103

Kaposi's sarcoma and HIV, 100
kissing and viruses, 97
Koch's postulates, 52, 145–147
Kuru, 50
    relatedness to scrapie, 50

mad cow disease, 42, 48–54
malaria and EBV, 98
male sex and viruses, 101–103
malnutrition and viral infection, 36
measles, 44–46
    origin of, 44
    relatedness to Hendra, 48
    relatedness to Nipah, 48
    relatedness to rinderpest, 44
    severity of, 45
    and vitamin A deficiency, 45
monkey pox in humans, 137–138
mosquitoes and viruses, 70, 71, 76
    urban mosquitoes, 71, 72

Nipah virus, bat origin of, 48

oral polio vaccine (OPV), 35
origin of life and viruses, 32–33, 156
outbreaks of viruses, 148

pest control by viruses, 135–137
phage therapy, 60
    cholera control by, 60, 61
    plague control by, 60
placenta formation and viruses, 92
plant viruses, 28
    and loss of crops, 28
poliovirus, 35–40
predator/prey survival and virus infection, 43
preserved foods and viruses, 99
prions, 50–52
Prusiner, Stanley, 54

quasi-species, 7
    Eigen's concept of, 7
quasi-species of viruses, 85
    and virus survival, 85

resistance
    resistance gene for HIV, 110–112

to viruses, 82–84, 110–112
of viruses to drugs, 85
retroviruses, 79
    as cause of cance, 80
    egg and milk transmission of, 81
    mother-to-child transmission of,
        81
rinderpest, 42–44
rodents and viruses, 76

Sabin, Albert, 35, 95
Salk, Jonas, 40
SARS
    animal reservoir of, 149, 150
    caused by coronavirus, 147
    China, and, 140
    protection against, 142
seasonality of viruses, 152
sex and viruses, 96
sexual transmission of virus. *See*
        transmission route of viruses
smallpox
    disease by, 131
    eradication of, 125, 132–134
    New World, and the, 130
    outbreaks of, 130, 135
    vaccination against, 132–134
Spanish flu, 16. *See also* influenza
        virus, 1918 epidemic of
species
    and extinction, 7, 125, 128
    forms of life, 6, 125
    relation to viruses, 6, 12
stress and virus infection, 69, 96

Temin, Howard, 79, 123
transmission routes of viruses, 101,
        102, 104

urbanization and viruses, 72

Vibrio cholerae. *See* cholera
viral aggression. *See* virus virulence
viral fitness, 8, 12
        champions of, 9

group vs. individual, 8, 137
    and virus survival, 8, 9, 11, 156
viral genes
    and host adaptation, 10
    and host protection, 10
    and placenta formation, 10
viral heart disorder, 35
    and Coxsackie virus, 35
viroids, 30–32
    and host diversity, 30
virulence of viruses, 136
    and virus survival, 158
virus to host help, 91
virus to virus help, 90
virus virulence and spread, 117
viruses
    and biodiversity, 65
    cancer and salted fish, 99
    and care for the host, 11
    designated "human", 128
    eternal life of, 7, 156
    extinction of, 126
    fitness of, 8, 12, 156
    genealogy of, 7
    genetic variants of, 7
    and harvests, 28
    life of, 4, 5, 126, 156
    and life on earth, 10, 11, 126
    and malnutrition, 28
    origin of, 156
    and the origin of mammals, 91–92
    and population growth, 12
    protection by, 91
    recombination of, 7
    survival of, 127, 156
viruses of primates. *See* jungle

water and viruses. *See* cholera
weather and viruses, 75, 76
West Nile, 67
    and bird migration, 69, 70
    origin of, 68

xenotransplantation, the risk of virus
        infection, 88, 89